iOS APPLICATION SECURITY

iOS APPLICATION SECURITY

The Definitive Guide for Hackers and Developers

by David Thiel

no starch press

San Francisco

Printed in USA

First printing

20 19 18 17 16   1 2 3 4 5 6 7 8 9

ISBN-10: 1-59327-601-X
ISBN-13: 978-1-59327-601-0

Publisher: William Pollock
Production Editor: Alison Law
Cover Illustration: Garry Booth
Interior Design: Octopod Studios
Developmental Editor: Jennifer Griffith-Delgado
Technical Reviewer: Alban Diquet
Copyeditor: Kim Wimpsett
Compositor: Alison Law
Proofreader: James Fraleigh

For information on distribution, translations, or bulk sales, please contact No Starch Press, Inc. directly:

No Starch Press, Inc.
245 8th Street, San Francisco, CA 94103
phone: 415.863.9900; info@nostarch.com
www.nostarch.com

Library of Congress Cataloging-in-Publication Data

```
Names: Thiel, David, 1980- author.
Title: iOS application security : the definitive guide for hackers and
    developers / by David Thiel.
Description: San Francisco : No Starch Press, [2016] | Includes index.
Identifiers: LCCN 2015035297| ISBN 9781593276010 | ISBN 159327601X
Subjects: LCSH: Mobile computing--Security measures. | iPhone
    (Smartphone)--Mobile apps--Security measures. | iPad (Computer)--Security
    measures. | iOS (Electronic resource) | Application software--Development.
    | Objective-C (Computer program language)
Classification: LCC QA76.9.A25 T474 2016 | DDC 004--dc23
LC record available at http://lccn.loc.gov/2015035297
```

To whomever I happen to be dating right now.

And to my parents, for attempting to restrict
my computer access as a child.

Also cats. They're pretty great.

About the Author

David Thiel has nearly 20 years of computer security experience. Thiel's research and book *Mobile Application Security* (McGraw-Hill) helped launch the field of iOS application security, and he has presented his work at security conferences like Black Hat and DEF CON. An application security consultant for years at iSEC Partners, Thiel now works for the Internet.org Connectivity Lab.

About the Technical Reviewer

Alban Diquet is a software engineer and security researcher who specializes in security protocols, data privacy, and mobile security, with a focus on iOS. Diquet has released several open source security tools, such as SSLyze, iOS SSL Kill Switch, and TrustKit. Diquet has also presented at various security conferences, including Black Hat, Hack in the Box, and Ruxcon.

BRIEF CONTENTS

Foreword by Alex Stamos xix
Acknowledgments xxi
Introduction xxiii

PART I: IOS FUNDAMENTALS

Chapter 1: The iOS Security Model 3
Chapter 2: Objective-C for the Lazy 13
Chapter 3: iOS Application Anatomy 27

PART II: SECURITY TESTING

Chapter 4: Building Your Test Platform 41
Chapter 5: Debugging with lldb and Friends 61
Chapter 6: Black-Box Testing 77

PART III: SECURITY QUIRKS OF THE COCOA API

Chapter 7: iOS Networking 107
Chapter 8: Interprocess Communication. 131
Chapter 9: iOS-Targeted Web Apps. 147
Chapter 10: Data Leakage 161
Chapter 11: Legacy Issues and Baggage from C 189
Chapter 12: Injection Attacks 199

PART IV: KEEPING DATA SAFE

Chapter 13: Encryption and Authentication 211
Chapter 14: Mobile Privacy Concerns 233

Index. 249

CONTENTS IN DETAIL

FOREWORD by Alex Stamos   xix

ACKNOWLEDGMENTS   xxi

INTRODUCTION   xxiii

Who This Book Is For xxiv
What's in This Book xxiv
    How This Book Is Structured xxiv
    Conventions This Book Follows xxvi
    A Note on Swift xxvi
Mobile Security Promises and Threats xxvii
    What Mobile Apps Shouldn't Be Able to Do xxvii
    Classifying Mobile Security Threats in This Book xxviii
Some Notes for iOS Security Testers xxx

PART I
IOS FUNDAMENTALS

1
THE IOS SECURITY MODEL   3

Secure Boot 4
Limiting Access with the App Sandbox 4
Data Protection and Full-Disk Encryption 5
    The Encryption Key Hierarchy 6
    The Keychain API 7
    The Data Protection API 7
Native Code Exploit Mitigations: ASLR, XN, and Friends 8
Jailbreak Detection 9
How Effective Is App Store Review? 10
    Bridging from WebKit 11
    Dynamic Patching 11
    Intentionally Vulnerable Code 12
    Embedded Interpreters 12
Closing Thoughts 12

2
OBJECTIVE-C FOR THE LAZY   **13**

Key iOS Programming Terminology 14
Passing Messages 14
Dissecting an Objective-C Program... 15
    Declaring an Interface 15
    Inside an Implementation File 16
Specifying Callbacks with Blocks.. 18
How Objective-C Manages Memory 18
Automatic Reference Counting 19
Delegates and Protocols... 20
    Should Messages 20
    Will Messages.. 20
    Did Messages 20
    Declaring and Conforming to Protocols 21
The Dangers of Categories... 22
Method Swizzling 23
Closing Thoughts 25

3
IOS APPLICATION ANATOMY   **27**

Dealing with plist Files 29
Device Directories 32
The Bundle Directory.. 33
The Data Directory 34
    The Documents and Inbox Directories 34
    The Library Directory 35
    The tmp Directory 37
The Shared Directory 37
Closing Thoughts 38

PART II
SECURITY TESTING

4
BUILDING YOUR TEST PLATFORM   **41**

Taking Off the Training Wheels 41
Suggested Testing Devices 42
Testing with a Device vs. Using a Simulator 43
Network and Proxy Setup 43
    Bypassing TLS Validation 44
    Bypassing SSL with stunnel... 46

Certificate Management on a Device 47
Proxy Setup on a Device... 48
Xcode and Build Setup.. 50
Make Life Difficult 51
Enabling Full ASLR 53
Clang and Static Analysis 54
Address Sanitizer and Dynamic Analysis 55
Monitoring Programs with Instruments 55
Activating Instruments 55
Watching Filesystem Activity with Watchdog 58
Closing Thoughts 59

5
DEBUGGING WITH LLDB AND FRIENDS 61

Useful Features in lldb 62
Working with Breakpoints 62
Navigating Frames and Variables 64
Visually Inspecting Objects.. 68
Manipulating Variables and Properties 69
Breakpoint Actions 70
Using lldb for Security Analysis 72
Fault Injection 72
Tracing Data 74
Examining Core Frameworks 74
Closing Thoughts 75

6
BLACK-BOX TESTING 77

Installing Third-Party Apps 78
Using a .app Directory 78
Using a .ipa Package File 80
Decrypting Binaries... 80
Launching the debugserver on the Device 81
Locating the Encrypted Segment 84
Dumping Application Memory 87
Reverse Engineering from Decrypted Binaries 89
Inspecting Binaries with otool 90
Obtaining Class Information with class-dump 92
Extracting Data from Running Programs with Cycript 93
Disassembly with Hopper 94
Defeating Certificate Pinning 96
Hooking with Cydia Substrate 97
Automating Hooking with Introspy 100
Closing Thoughts 103

PART III
SECURITY QUIRKS OF THE COCOA API

7
IOS NETWORKING 107

Using the iOS URL Loading System 108
    Using Transport Layer Security Correctly 108
    Basic Authentication with NSURLConnection 110
    Implementing TLS Mutual Authentication with NSURLConnection 112
    Modifying Redirect Behavior 113
    TLS Certificate Pinning 114
Using NSURLSession 117
    NSURLSession Configuration 117
    Performing NSURLSession Tasks 118
    Spotting NSURLSession TLS Bypasses 119
    Basic Authentication with NSURLSession 119
    Managing Stored URL Credentials 121
Risks of Third-Party Networking APIs 122
    Bad and Good Uses of AFNetworking 122
    Unsafe Uses of ASIHTTPRequest 124
Multipeer Connectivity 125
Lower-Level Networking with NSStream 127
Even Lower-level Networking with CFStream 128
Closing Thoughts 129

8
INTERPROCESS COMMUNICATION 131

URL Schemes and the openURL Method 132
    Defining URL Schemes 132
    Sending and Receiving URL/IPC Requests 133
    Validating URLs and Authenticating the Sender 134
    URL Scheme Hijacking 136
Universal Links 137
Sharing Data with UIActivity 139
Application Extensions 140
    Checking Whether an App Implements Extensions 141
    Restricting and Validating Shareable Data 142
    Preventing Apps from Interacting with Extensions 143
A Failed IPC Hack: The Pasteboard 144
Closing Thoughts 145

9
IOS-TARGETED WEB APPS   147

Using (and Abusing) UIWebViews 147
     Working with UIWebViews 148
     Executing JavaScript in UIWebViews 149
Rewards and Risks of JavaScript-Cocoa Bridges 150
     Interfacing Apps with JavaScriptCore 150
     Executing JavaScript with Cordova 154
Enter WKWebView 158
     Working with WKWebViews 158
     Security Benefits of WKWebViews 159
Closing Thoughts 160

10
DATA LEAKAGE   161

The Truth About NSLog and the Apple System Log 161
     Disabling NSLog in Release Builds 163
     Logging with Breakpoint Actions Instead 164
How Sensitive Data Leaks Through Pasteboards 164
     Restriction-Free System Pasteboards 165
     The Risks of Custom-Named Pasteboards 165
     Pasteboard Data Protection Strategies 167
Finding and Plugging HTTP Cache Leaks 169
     Cache Management 170
     Solutions for Removing Cached Data 171
     Data Leakage from HTTP Local Storage and Databases 174
Keylogging and the Autocorrection Database 175
Misusing User Preferences 178
Dealing with Sensitive Data in Snapshots 178
     Screen Sanitization Strategies 179
     Why Do Those Screen Sanitization Strategies Work? 182
     Common Sanitization Mistakes 183
     Avoiding Snapshots by Preventing Suspension 183
Leaks Due to State Preservation 184
Secure State Preservation 185
Getting Off iCloud to Avoid Leaks 187
Closing Thoughts 188

11
LEGACY ISSUES AND BAGGAGE FROM C   189

Format Strings 190
     Preventing Classic C Format String Attacks 191
     Preventing Objective-C Format String Attacks 192

Buffer Overflows and the Stack 193
    A strcpy Buffer Overflow... 194
    Preventing Buffer Overflows 195
Integer Overflows and the Heap 196
    A malloc Integer Overflow 197
    Preventing Integer Overflows 198
Closing Thoughts 198

12
INJECTION ATTACKS   199

Client-Side Cross-Site Scripting.. 199
    Input Sanitization 200
    Output Encoding 201
SQL Injection 203
Predicate Injection 204
XML Injection 205
    Injection Through XML External Entities 205
    Issues with Alternative XML Libraries 207
Closing Thoughts 207

PART IV
KEEPING DATA SAFE

13
ENCRYPTION AND AUTHENTICATION   211

Using the Keychain 211
    The Keychain in User Backups 212
    Keychain Protection Attributes 212
    Basic Keychain Usage 214
    Keychain Wrappers 217
    Shared Keychains.. 218
    iCloud Synchronization 219
The Data Protection API 219
    Protection Levels 220
    The DataProtectionClass Entitlement.. 223
    Checking for Protected Data Availability 224
Encryption with CommonCrypto 225
    Broken Algorithms to Avoid 226
    Broken Initialization Vectors... 226
    Broken Entropy 227
    Poor Quality Keys... 227
Performing Hashing Operations... 228
Ensuring Message Authenticity with HMACs 229
Wrapping CommonCrypto with RNCryptor 230

Local Authentication: Using the TouchID 231
      How Safe Are Fingerprints? 232
Closing Thoughts 232

14
MOBILE PRIVACY CONCERNS   233

Dangers of Unique Device Identifiers 233
      Solutions from Apple 234
      Rules for Working with Unique Identifiers................................... 235
Mobile Safari and the Do Not Track Header 236
Cookie Acceptance Policy... 237
Monitoring Location and Movement 238
      How Geolocation Works 238
      The Risks of Storing Location Data 238
      Restricting Location Accuracy 239
      Requesting Location Data 240
Managing Health and Motion Information 240
      Reading and Writing Data from HealthKit 241
      The M7 Motion Processor... 242
Requesting Permission to Collect Data 243
Proximity Tracking with iBeacons 244
      Monitoring for iBeacons 244
      Turning an iOS Device into an iBeacon 246
      iBeacon Considerations 247
Establishing Privacy Policies 247
Closing Thoughts 248

INDEX   249

FOREWORD

Prior to the digital age, people did not typically carry a cache of sensitive personal information with them as they went about their day. Now it is the person who is not carrying a cell phone, with all that it contains, who is the exception. . . .

Modern cell phones are not just another technological convenience. With all they contain and all they may reveal, they hold for many Americans "the privacies of life". . . . The fact that technology now allows an individual to carry such information in his hand does not make the information any less worthy of the protection for which the Founders fought.

— Chief Justice John Roberts, Riley v. California (2014)

Few would argue that the smartphone has been, by far, the most impactful technological advance of the 21st century. Since the release of the iPhone in 2007, the number of active smartphones has skyrocketed. As I write this at the end of 2015, there are nearly 3.4 billion in use; that's one for just about half the human population (somewhere over 7.3 billion). Globally, phones have easily eclipsed all other types of computers used to access the Internet, and an entire book could be filled with examples of how near-ubiquitous access is shaping human civilization. Mobile is changing the world, and has enriched countless lives by bringing widespread access to educational resources, entertainment, and unprecedented economic opportunities. In some parts of the world, mobile connectivity and social networking has even led to the downfall of autocratic regimes and the realignment of societies.

Even the septuagenarians on the US Supreme Court have recognized the power of modern mobile computing, setting new legal precedents with judgements, like Riley v. California quoted above, that recognize that a smartphone is more than just a device—it is a portal into the private aspects of everyone's lives.

Like all technological revolutions, the mobile revolution has its downsides. Our ability to connect with the far side of the world does nothing to improve the way we communicate with those in front of our faces, and mobile has done nothing to eliminate the world's long-established economic disparities. At the same time, as with enterprise computing, personal computing, and networking revolutions, smartphones have introduced new kinds of potential security flaws, and introduced or reinvented all kinds of security and safety issues.

While the proto-smartphones released prior to 2007 brought us several important technological innovations, it was the subsequent publishing of rich SDKs and the opening of centralized app stores that turned the new mobile computers into platforms for third-party innovation. They also created a whole new generation of developers who now need to adapt the security lessons of the past to a new, uncertain threat landscape.

In the ten years I have known David Thiel, I have constantly been impressed by his desire to examine, disassemble, break, and understand the latest technologies and apply his knowledge to improving the security of others. David was one of the first people to recognize the fascinating security challenges and awesome potential of the iPhone, and since the first days of what was then the iPhone OS SDK, he has studied the ways app developers could stumble and expose their users to risk, or rise above the limitations of the platform to build privacy- and safety-enhancing applications.

This book contains the most thorough and thoughtful treatment of iOS security that you can find today. Any iOS developer who cares about their customers should use it to guide their product, architecture, and engineering decisions and to learn from the mistakes that David has spent his career finding and fixing.

The smartphone revolution has tremendous potential, but only if we do the utmost to protect the safety, trust, and privacy of the people holding these devices, who want to enrich their lives through our inventions.

Alex Stamos
Chief Security Officer, Facebook

ACKNOWLEDGMENTS

Thanks to Jennifer Griffith-Delgado, Alison Law, Bill Pollock, and the rest of the No Starch team, as well as Tom Daniels for his major contributions to Chapter 9, and Alban Diquet and Chris Palmer for their excellent review and feedback.

INTRODUCTION

Much has been written regarding iOS's security model, jailbreaking, finding code execution vulnerabilities in the base OS, and other security-related characteristics. Other work has focused on examining iOS from a forensic perspective, including how to extract data from physical devices or backups as part of criminal investigations. That information is all useful, but this book aims to fill the biggest gaps in the iOS literature: applications.

Little public attention has been given to actually writing secure applications for iOS or for performing security evaluations of iOS applications. As a consequence, embarrassing security flaws in iOS applications have allowed for exposure of sensitive data, circumvention of authentication mechanisms, and abuse of user privacy (both intentional and accidental). People are using iOS applications for more and more crucial tasks and entrusting them with a lot of sensitive information, and iOS application security needs to mature in response.

As such, my goal is for this book is to be as close as possible to the canonical work on the secure development of iOS applications in particular. iOS is a rapidly moving target, of course, but I've tried to make things as accurate as possible and give you the tools to inspect and adapt to future API changes.

Different versions of iOS also have different flaws. Since Apple has "end-of-lifed" certain devices that developers may still want their applications to run on (like the iPad 1), this book covers flaws present in iOS versions 5.*x* to 9.0 (the latest at the time of writing) and, where applicable, discusses risks and mitigations specific to each version.

Who This Book Is For

First, this is a book about security. If you're a developer or security specialist looking for a guide to the common ways iOS applications fail at protecting their users (and the options available to you or a client for patching those holes), you're in the right place.

You'll get the most out of this book if you have at least a little experience with iOS development or a passing familiarity with how iOS applications work under the hood. But even without that knowledge, as long as you're an experienced programmer or penetration tester who's not afraid to dig in to Apple's documentation as needed, you should be fine. I give a whirlwind tour of Objective-C and its most commonly used API, Cocoa Touch, in Chapter 2, so if you need some high-level basics or a refresher on the language, start there.

What's in This Book

I've been performing a wide variety of iOS application security reviews and penetration tests since about 2008, and I've collected a lot of knowledge on the pitfalls and mistakes real-world developers encounter when writing iOS applications. This book boils down that knowledge to appeal both to iOS developers looking to learn the practice of secure development and to security specialists wanting to learn how to spot problems in iOS security.

How This Book Is Structured

In **Part I: iOS Fundamentals**, you'll dig in to the background of iOS, its security history, and its basic application structure.

- **Chapter 1: The iOS Security Model** briefly examines the iOS security model to give you an idea of the platform's fundamental security protections and what they can and cannot provide.

- **Chapter 2: Objective-C for the Lazy** explains how Objective-C differs from other programming languages and gives a quick overview of its terminology and design patterns. For seasoned Objective-C programmers, this may not be new information, but it should be valuable to beginners and others dabbling in iOS for the first time.

- **Chapter 3: iOS Application Anatomy** outlines how iOS applications are structured and bundled and investigates the local storage mechanisms that can leak sensitive information.

In **Part II: Security Testing**, you'll see how to set up your security testing environment, for use either in development or in penetration testing. I'll also share some tips for setting up your Xcode projects to get the most out of the available security mechanisms.

- **Chapter 4: Building Your Test Platform** gives you all the information that you need to get started with tools and configurations to help you audit and test iOS applications. This includes information on using the Simulator, configuring proxies, bypassing TLS validation, and analyzing application behavior.

- **Chapter 5: Debugging with lldb and Friends** goes deeper into monitoring application behavior and bending it to your will using lldb and Xcode's built-in tools. This will help you analyze more complex problems in your code, as well as give you a test harness to do things like fault injection.

- **Chapter 6: Black-Box Testing** delves into the tools and techniques that you'll need to successfully analyze applications that you don't have source code for. This includes basic reverse engineering, binary modification, copying programs around, and debugging on the device with a remote instance of lldb.

In **Part III: Security Quirks of the Cocoa API**, you'll look at common security pitfalls in the Cocoa Touch API.

- **Chapter 7: iOS Networking** discusses how networking and Transport Layer Security work in iOS, including information on authentication, certificate pinning, and mistakes in TLS connection handling.

- **Chapter 8: Interprocess Communication** covers interprocess communication mechanisms, including URL schemes and the newer Universal Links mechanism.

- **Chapter 9: iOS-Targeted Web Apps** covers how web applications are integrated with iOS native apps, including working with web views or using JavaScript/Cocoa bridges such as Cordova.

- **Chapter 10: Data Leakage** discusses the myriad ways that sensitive data can unintentionally leak onto local storage, to other applications, or over the network.

- **Chapter 11: Legacy Issues and Baggage from C** gives an overview of C flaws that persist in iOS applications: stack and heap corruption, format string flaws, use-after-free, and some Objective-C variants of these classic flaws.

- **Chapter 12: Injection Attacks** covers attacks such as SQL injection, cross-site scripting, XML injection, and predicate injection, as they relate to iOS applications.

Finally, **Part IV: Keeping Data Safe** covers issues relating to privacy and encryption.

- **Chapter 13: Encryption and Authentication** looks at encryption best practices, including how to properly use the Keychain, the Data Protection API, and other cryptographic primitives provided by the CommonCrypto framework.

- **Chapter 14: Mobile Privacy Concerns** ends the book with a discussion of user privacy, including what collecting more data than needed can mean for both application creators and users.

By the end of this book, you should be well equipped to grab an application, with or without source code, and quickly pinpoint security bugs. You should also be able to write safe and secure applications for use in the wider world.

Conventions This Book Follows

Because Objective-C is a rather verbose language with many extremely long class and method names, I've wrapped lines in source code listings to maximize clarity. This may not reflect the way you'd actually want to format your code. In some cases, the results are unavoidably ugly—if wrapping makes the code seem less clear, try pasting it into Xcode and allowing Xcode to reformat it.

As I will detail in Chapter 2, I favor the traditional Objective-C infix notation instead of dot notation. I also put curly braces on the same line as method declarations for similar reasons: I'm old.

Objective-C class and method names will appear in monospaced font. C functions will appear in monospaced font as well. For brevity and cleanliness, the path */Users/<your username>/Library/Developer/CoreSimulator/* will be referred to as *$SIMPATH.*

A Note on Swift

There's been much interest in the relatively new Swift language, but you'll find I don't cover it in this book. There are a few reasons why.

First, I have yet to actually come across a production application written in Swift. Objective-C is still far and away the most popular language for iOS applications, and we'll be dealing with code written in it for many years to come.

Second, Swift just has fewer problems. Since it's not based on C, it's easier to write safer code, and it doesn't introduce any new security flaws (as far as anyone knows).

Third, because Swift uses the same APIs as Objective-C, the security pitfalls in the Cocoa Touch API that you may run into will be basically the same in either language. The things you learn in this book will almost all apply to both Objective-C and Swift.

Also, Swift doesn't use infix notation and square brackets, which makes me sad and confused.

Mobile Security Promises and Threats

When I first started working with mobile applications, I honestly questioned the need for a separate mobile application security category. I considered mobile applications to be the same as desktop applications when it came to bugs: stack and heap overflows, format string bugs, use-after-free, and other code execution issues. While these are still possible in iOS, the security focus for mobile devices has expanded to include privacy, data theft, and malicious interprocess communication.

As you read about the iOS security specifics I cover in this book, keep in mind that users expect apps to avoid doing certain things that will put their security at risk. Even if an app avoids overtly risky behaviors, there are still several threats to consider as you fortify that app's defenses. This section discusses both security promises an app makes to its users and the types of attacks that can force an app to break them.

What Mobile Apps Shouldn't Be Able to Do

Learning from the design mistakes of earlier desktop operating systems, the major mobile operating systems were designed with application segregation in mind. This is different from desktop applications, where any application a user runs more or less has access to all that user's data, if not control of the entire machine.

As a result of increased focus on segregation and general improvements in the mobile OS arena, user expectations have expanded. In general, mobile applications (including yours) should be unable to do a few key things.

Cause Another Application to Misbehave

Applications shouldn't be able to crash or meddle with other applications. In the bad old days, not only could other applications generally read, modify, or destroy data, they could take down the entire OS with that data. As time went on, desktop process segregation improved but primarily with the goal of increasing stability, rather than addressing security or privacy concerns.

Mobile operating systems improve upon this, but total process segregation is not possible while fulfilling users' interoperability needs. The boundary between applications will always be somewhat porous. It's up to developers to ensure that their applications don't misbehave and to take all prudent measures to safeguard data and prevent interference from malicious applications.

Deny Service to a User

Given that iOS has historically been used primarily on phones, it's crucial that an application not be able to do something that would prevent the user from making an emergency call. In many places, this is a legal requirement, and it's the reason for protective measures that keep attackers (and users) from tampering with the underlying OS.

Steal a User's Data

An application should not be able to read data from other applications or the base OS and deliver it to a third party. It should also not be able to access sensitive user data without the permission of the user. The OS should keep applications from reading data directly from other application's data stores, but preventing theft via other channels requires developers to pay attention to what IPC mechanisms an application sends or receives data on.

Cost the User Unexpected Money

Apps shouldn't be able to incur charges without the user's approval. Much of the mobile malware that has been found in the wild has used the ability to send SMS messages to subscribe the user to third-party services, which pass charges through to the user's phone provider. Purchases made within the application should be clear to the user and require explicit approval.

Classifying Mobile Security Threats in This Book

To help understand mobile device security threats and their mitigations, it's also useful to keep a few attack types in mind. This keeps our analysis of threats realistic and helps to analyze the true impact of various attacks and their defenses.

Forensic Attacks

Forensic attackers come into possession of a device or its backups, intending to extract its secrets. Most often, this involves examination of the physical storage on the device. Because phone or tablet theft is relatively easy and common compared to stealing other computing devices, much more attention is placed on forensics.

Forensic attacks can be performed by either an opportunistic attacker or a skilled attacker targeting a specific individual. For opportunistic attackers, extracting information can be as simple as stealing a phone without any PIN protection; this allows them to steal images, notes, and any other data normally accessible on the phone. It can also assist an attacker in compromising services that use two-factor authentication in conjunction with a phone-based token or SMS.

A skilled forensic attacker could be a rogue employee, corporation, government, law enforcement official, or perhaps really motivated extortionist. This kind of attacker knows the techniques to perform a temporary jailbreak, crack simple PINs, and examine data throughout the device's filesystem, including system-level and application-level data. This can provide an attacker with not just data presented through the UI but the underlying cache information, which can include screenshots, keystrokes, sensitive information cached in web requests, and so forth.

I'll cover much of the data of interest to forensic attackers in Chapter 10, as well as some further protective measures in Chapter 13.

Code Execution Attacks

Remote code execution attacks involve compromising the device or its data by execution of code on the device, without having physical possession of the device. This can happen via many different channels: the network, QR codes or NFC, parsing of maliciously crafted files, or even hostile hardware peripherals. Note that after gaining code execution on a device, many of the forensic attacks used to expose user secrets are now possible. There are a few basic subtypes of code execution attacks that frequently result from lower-level programming flaws, which I'll discuss in Chapter 11.

Web-Based Attacks

Web-based remote code execution attacks primarily use maliciously crafted HTML and JavaScript to mislead the user or steal data. A remote attacker either operates a malicious website, has taken over a legitimate website, or simply posts maliciously crafted content to a public forum.

These attacks can be used to steal data from local data stores such as HTML5 database storage or localStorage, alter or steal data stored in SQLite databases, read session cookies, or plant a fake login form to steal a user's credentials. I'll talk more about web application–related issues in Chapter 9 and Chapter 12.

Network-Based Attacks

Network-based code execution attacks attempt to gain control over an application or the entire system by injecting executable code of some type over the network. This can be either modification of network traffic coming into the device or exploitation of a system service or the kernel with a code execution exploit. If the exploit targets a process with a high degree of privilege, the attacker can gain access not only to the data of a specific application but to data all over the device's storage. They can also monitor the device's activity and plant backdoors that will allow later access. I'll talk specifically about network-related APIs in Chapter 7.

Attacks That Rely on Physical Proximity

Physical code execution attacks tend to be exploits that target devices using communications such as NFC or the USB interface. These types of attacks have been used for jailbreaking in the past but can also be used to compromise the device using brief physical interaction. Many of these attacks are on the OS itself, but I'll discuss some issues relating to physical proximity in Chapter 14.

Some Notes for iOS Security Testers

It's my strong belief that penetration tests should be performed with source code if at all possible. While this is not representative of the position of most external attackers, it does maximize the ability to find important bugs within a limited time frame. Real-world attackers have as much time as they care to spend on analyzing your application, and Objective-C lends well to reverse engineering. They'll figure it out, given the time. However, most penetration tests are limited by time and money, so simulating a real-world attacker should not usually be the goal.

I cover both white-box (that is, source-assisted) and black-box methodologies in this book, but the focus will be on source-assisted penetration tests because this finds more bugs faster and helps with learning the standard Cocoa library. Many techniques I describe in this book lend well to either approach.

All that said, iOS developers come from many different disciplines, and each person's skill set affects the types of security issues that slip into an app unnoticed. Whether you're testing someone else's application or trying to poke holes in your own, keep in mind a few different development backgrounds as you test.

Some iOS developers come from a C or C++ background, and since we all tend to use what we know, you'll find their codebases often use C/C++ APIs rather than Cocoa equivalents. If you know an application under test was created by former C/C++ programmers, you may find Chapter 11 to be useful reading because it discusses issues commonly found in straight C/C++ code.

For some new programmers, Objective-C is actually their first programming language. They often haven't learned that many vanilla C APIs, so ideally, you'll find fewer of those issues. There's also the rare wizened NeXTStep programmer who's made the move to OS X or iOS, with a library of collected wisdom regarding NeXTStep/Cocoa APIs but less mobile experience. If either sounds like you or your client, you'll find the chapters in Part III most helpful.

Programmers with Java backgrounds might try to force Java design patterns onto an application, endlessly abstracting functionality. Web developers who have been drafted into writing a mobile application, on the other hand, may try to wrap as much code as possible into a web app, writing minimal applications that rely on WebKit to view application content. Check out Chapter 9 for some WebKit-related pitfalls.

Developers with the last few skill sets I mentioned are less likely to use low-level APIs, which can prevent classic C flaws. They are, however, unlikely to spot mistakes when using those low-level APIs, so you'll want to pay close attention if they use them.

Of course, none of these backgrounds is necessarily better suited to secure development than the others—both high-level and low-level APIs can be abused. But when you know how existing skills can affect the writing of iOS applications, you're a step closer to finding and solving security issues.

My own background is that of a penetration tester, which I consider akin to being an art critic: I *can* write code, but the vast majority of my time is spent looking at other people's code and telling them what's wrong with it. And like in the art world, the majority of that code is rather crap. Unlike the art world, however, code problems can often be fixed with a patch. My hope is that at the end of this book, you'll be able to spot bad iOS code and know how to start plugging the holes.

PART I

IOS FUNDAMENTALS

1

THE IOS SECURITY MODEL

Let's give credit where credit is due: Apple has been pretty successful in keeping malicious software out of the App Store (as far as I know). But the application review process can be a frustrating black box for developers. The process used by Apple's reviewers is not publicly documented, and sometimes it's simply not clear what functionality is and isn't permitted. Apple gives some decent guidelines,[1] but apps have been rejected based on criteria that apply to accepted applications as well.

Of course, what qualifies as malicious is defined by Apple, not by users. Apple uses the App Store as a way to control what functionality is available on the iOS platform, meaning the only way to obtain certain functionality is to jailbreak the device or subvert the App Store review process. An example of this is the Handy Light application, which masqueraded as a flashlight application but contained a hidden mode to enable device tethering.[2]

---

1. *https://developer.apple.com/appstore/resources/approval/guidelines.html*
2. *http://www.macworld.com/article/1152835/iphone_flashlight_tethering.html*

The app review process on its own will never catch all sophisticated (or trivial) malicious applications, so other mechanisms are needed to effectively keep bad applications from affecting the wider OS environment. In this chapter, you'll learn about the architecture of iOS's security mechanisms; in later chapters, you'll dig in to how to take advantage of these mechanisms properly in your own programs.

Let's take a quick look at the fundamental security components iOS implements to prevent exploits and protect data. I'll dive deeper into the actual mechanics of most of these in later sections, but I'll start by giving a broad overview of the impetus behind them and their utility.

Secure Boot

When you power on an iOS device, it reads its initial instructions from the read-only Boot ROM, which bootstraps the system. The Boot ROM, which also contains the public key of Apple's certificate authority, then verifies that the low-level bootloader (LLB) has been signed by Apple and launches it. The LLB performs a few basic tasks and then verifies the second-stage bootloader, iBoot. When iBoot launches, the device can either go into recovery mode or boot the kernel. After iBoot verifies the kernel is also signed by Apple, the boot process begins in earnest: drivers are loaded, devices are probed, and system daemons start.

The purpose of this chain of trust is to ensure that all components of the system are written, signed, and distributed by Apple—not by third parties, which could include malicious attackers and authors of software intended to run on jailbroken devices. The chain is also used to bootstrap the signature checking of individual applications; all applications must be directly or indirectly signed by Apple.

Attacking this chain of trust is how jailbreaking works. Jailbreak authors need to find a bug somewhere in this chain to disable the verification of the components further down the chain. Exploits of the Boot ROM are the most desirable because this is the one component Apple can't change in a software update.

Limiting Access with the App Sandbox

Apple's sandbox, historically referred to as Seatbelt, is a *mandatory access control (MAC)* mechanism based on FreeBSD's TrustedBSD framework, primarily driven by Robert Watson. It uses a Lisp-like configuration language to describe what resources a program can or cannot access, including files, OS services, network and memory resources, and so on.

MAC is different from traditional access control mechanisms such as discretionary access control (DAC) in that it disallows *subjects*, such as user processes, from manipulating the access controls on *objects* (files, sockets, and so on). DAC, in its simplest, most common form, is controlled on a UNIX system with *user, group,* and *other* permissions, all of which can be granted read, write, or execute permissions.[3] In a DAC system, users can change permissions if they have ownership of an object. For example, if you own a file, you can set it to be world-readable or world-writable, which obviously subverts access controls.

While MAC is a broad term, in sandbox-land it means that applications are shunted into a virtual container that consists of detailed rules specifying which system resources a subject is allowed to access, such as network resources, file read and writes, the ability to fork processes, and so on.[4] On OS X you can control some of how your application is sandboxed, but on iOS all third-party applications are run with a single restrictive policy.

In terms of file access, processes are generally confined to their own application bundle directory; they can read and write only the files stored there. The standard policy is slightly porous, however. For example, in some versions of iOS, photos in */private/var/mobile/Media/Photos/* can be directly accessed by third-party applications, despite being outside the application's bundle directory, which allows programs to surreptitiously access photos without asking for user permission. The only protection against applications abusing this type of privilege is Apple's application review process.

This approach differs from that used by Android, which implements a more traditional DAC model, where applications are given their own user ID and a directory owned by that ID. Permissions are managed strictly via traditional UNIX file permissions. While both approaches are workable, MAC generally provides more flexibility. For instance, in addition to app directory segregation, MAC policies can be used to restrict network access or limit what actions system daemons can take.

Data Protection and Full-Disk Encryption

iOS led the way in offering mobile devices with filesystem encryption, for which some credit is due. iOS offers full-disk encryption and additionally provides developers with the Data Protection API to further protect their files. These two related mechanisms make it possible to wipe remote devices and protect user data in the event of device theft or compromise.

Historically, full-disk encryption is made to solve one problem: data at rest being stolen by an attacker. In the laptop or desktop world, this would involve either removing the hard drive from a machine and mounting it on a separate machine or booting into an OS that could read the files off the drive. Filesystem encryption does *not* protect against data being stolen off of a running device. If an application is able to read a file from the disk,

---

3. This description is, of course, slightly simplified; there are also sticky bits, setuid bits, and so forth. Since iOS doesn't use DAC as its primary access control mechanism, though, I won't get into those topics in this book.

4. You can find a good summary of the default iOS sandbox policies at *https://media.blackhat. com/bh-us-11/DaiZovi/BH_US_11_DaiZovi_iOS_Security_WP.pdf*

filesystem encryption provides no benefit because the kernel transparently decrypts files for any process that tries to read them. In other words, filesystem encryption works at a lower level than the calls typically used to read files. An attacker who can authenticate to the system can read any available files unimpeded.

iOS devices are generally designed to be running at all times, and their internal storage is not easily removable. If an attacker wanted to read sensitive data from a device without authenticating, they would have to completely disassemble the device and hook up the flash storage to a custom interface to read storage directly. There are several far easier methods for obtaining data from the device—including code execution exploits, jailbreaking, and so on—so no one would ever actually go to all that trouble.

But that doesn't mean iOS's full filesystem encryption is completely useless. It's necessary to correctly implement two other critical security features: secure file deletion and remote device wipe. Traditional methods of securely erasing files don't apply to iOS devices, which use solid-state drives (SSDs). The wear-reduction mechanisms used by this hardware remove all guarantees that overwriting a file actually overwrites the previous physical location of the file. The solution to this problem is to ensure that files are encrypted with safely stored keys so that in the event that data destruction is requested, keys can be discarded. The encryption key hierarchy used in iOS is layered. Entire classes of data or even the whole filesystem can be destroyed by throwing away a single encryption key.

The Encryption Key Hierarchy

Filesystem encryption keys for stored data on iOS are hierarchical, with keys encrypting other keys, so that Apple has granular control if and when data is available. The basic hierarchy is shown in Figure 1-1.

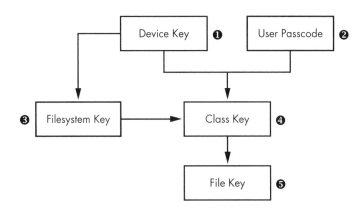

Figure 1-1: The simplified iOS encryption key hierarchy

The *File Key* ❺ is an individual key generated per file and stored in the file's metadata. The *Class Key* ❹ is a dedicated key for a particular Data Protection class so that files classified with different protection levels use separate cryptographic keys. In older versions of iOS, the default protection class was NSFileProtectionNone; from version 5 onward, the default protection class is NSFileProtectionCompleteUntilFirstUserAuthentication, which is further described in Chapter 13. The *Filesystem Key* ❸ is a global encryption key used to encrypt the file's security-related metadata after the metadata is encrypted by the Class Key.

The *Device Key* ❶, also known as the UID key, is unique for each device and accessible only by the hardware AES engine, not by the OS itself. This is the master key of the system, as it were, which encrypts the Filesystem Key and the Class Keys. The *User Passcode* ❷, if enabled, is combined with the Device Key when encrypting Class Keys.

When a passcode is set, this key hierarchy also allows developers to specify how they want their locally stored data to be protected, including whether it can be accessed while the device is locked, whether data gets backed up to other devices, and so on. You'll learn more about how to use encryption and file protection features to protect files from device thieves in Chapter 13, where I cover the Data Protection API in greater depth.

The Keychain API

For small pieces of secret information, iOS offers a dedicated Keychain API. This allows developers to store information such as passwords, encryption keys, and sensitive user data in a secure location not accessible to other applications. Calls to the Keychain API are mediated through the securityd daemon, which extracts the data from a SQLite data store. The programmer can specify under what circumstances keys should be readable by applications, similar to the Data Protection API.

The Data Protection API

The Data Protection API leverages filesystem encryption, the Keychain, and the user's passcode to provide an additional layer of protection to files at the developer's discretion. This limits the circumstances under which processes on the system can read such files. This API is most commonly used to make data inaccessible when a device is locked.

The degree of data protection in effect depends heavily on the version of iOS the device is running because the default Data Protection classes have changed over time. In newly created iOS application projects, Data Protection is enabled by default for all application data until the user unlocks the device for the first time after boot. Data Protection is enabled in project settings, as shown in Figure 1-2.

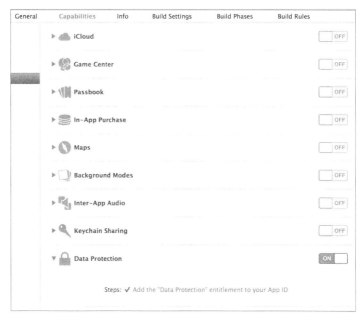

Figure 1-2: Adding a data protection entitlement to a project

Native Code Exploit Mitigations: ASLR, XN, and Friends

iOS implements two standard mechanisms to help prevent code execution attacks: *address space layout randomization (ASLR)* and the *XN bit* (which stands for *eXecute Never*). ASLR randomizes the memory location of the program executable, program data, heap, and stack on every execution of the program; because shared libraries need to stay put to be shared by multiple processes, the addresses of shared libraries are randomized every time the OS boots instead of every program invocation. This makes the specific memory addresses of functions and libraries hard to predict, preventing attacks such as a return-to-libc attack, which relies on knowing the memory addresses of basic libc functions. I'll talk more about these types of attacks and how they work in Chapter 11.

The XN bit, generally known on non-ARM platforms as the NX (No-eXecute) bit, allows the OS to mark segments of memory as nonexecutable, which is enforced by the CPU. In iOS, this bit is applied to a program's stack and heap by default. This means in the event that an attacker is able to insert malicious code onto the stack or heap, they won't be able to redirect the program to execute their attack code. Figure 1-3 shows the segments of process memory and their XN status.

A program can have memory that is both writable and executable only if it's signed with Apple's own code-signing entitlement; this is primarily used for the JavaScript just-in-time (JIT) compiler included as part of Mobile Safari. The regular WebViews that you can use in your own programs don't have access to the same functionality; this is to help prevent code execution attacks. An unfortunate effect of Apple's policy is that it effectively bans third-party JITs, notably preventing Chrome from performing as well as Safari on iOS. Chrome has to use the built-in WebViews.

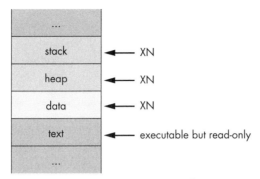

Figure 1-3: Basic memory segments of a process

Jailbreak Detection

Fundamentally, *jailbreaking* is any procedure that disables iOS's code-signing mechanisms, allowing a device to run applications other than those approved directly by Apple. Jailbreaking allows you to take advantage of some useful development and testing tools, as well as utilities that would never pass App Store muster.[5] The ability to jailbreak is critical to testing applications in a black-box fashion; I'll dig in to black-box testing further in Chapter 6.

Contrary to popular belief, jailbreaking doesn't necessarily disable the iOS sandbox. It just allows you to install applications outside of the sandbox. Applications installed in the home directory of the *mobile* user (that is, ones installed via the App Store) are still subject to sandbox restrictions. Third-party iOS applications that need higher levels of privilege on jailbroken devices are installed in the */Applications* folder, alongside the stock Apple applications.

The history of jailbreak detection is long and comical. This procedure is intended to detect whether the device is at heightened risk for compromise because of the less trustworthy nature of unsigned third-party programs. To be fair, there isn't a shortage of malware and misbehaving programs in third-party application repositories, but in general, jailbreak detection isn't worth your time because it won't stop a determined attacker.

For a brief period, Apple had an official jailbreak detection API, but this was pulled rather quickly from subsequent releases of iOS. In the absence of this API, developers have implemented a number of tricks to try detect jail-breaking themselves. The most popular techniques for attempting jailbreak detection go along these lines:

---

5. It seems, however, that most jailbreak users are motivated by the ability to perform the digital equivalent of putting spinning hubcaps on your car.

- Spawn a new process, such as using `fork()`, `vfork()`, `popen()`, and so on. This is something explicitly prevented by the sandbox. Of course, on jailbroken devices the sandbox is still enabled, making this strategy fairly pointless. It will fail for any App Store application regardless of whether the device is jailbroken.

- Read a file path outside of the sandbox. Developers commonly try to access the binary for `ssh`, `bash`, the *Cydia.app* directory, the path to the apt repository that Cydia uses, and so forth. These checks are painfully easy to get around, and tools such as Xcon[6] help end users bypass these checks automatically.

- Ensure that the method name with your jailbreak detection logic is something blatantly obvious, such as `isJailbroken`, allowing attackers to easily spot and disable your jailbreak checks.

There are some more obscure techniques as well. For example, Apple's iBooks application attempts to run unsigned code distributed with the app bundle.[7] Savvy developers will also attempt to use `_dyld_image_count()` and `_dyld_get_image_name()` to check the total number of loaded dynamic libraries (*dylibs*) and their names,[8] and use `_dyld_get_image_header()` to inspect their locations in memory.[9] Getting around these checks generally requires patching an application's binary directly.

As you may have noticed, I don't think much of jailbreak detection in general. Like binary obfuscation techniques and DRM, jailbreak detection techniques typically serve only to make you look foolish when they're bypassed (and believe me, I've seen some foolish obfuscation techniques). Proponents often argue that performing cursory jailbreak detection slows down pirates or attackers. But your adversary's hobby is cracking applications, and they have weeks of time on hand in which to do so—slowing them down by a few hours isn't really worthwhile. The longest it has taken me to develop a jailbreak detection bypass is about a day with an obfuscated binary and a battery of tests—and I'm an amateur at such things.

How Effective Is App Store Review?

When developing an application or assessing the threats that an app faces, it's important to evaluate the risk of a rogue application ending up on end users' devices. Any malicious third-party applications that make it onto devices are able to interact with applications via IPC mechanisms, as well as steal personal information. The primary defense against these applications is Apple's App Store review process.

---

6. *http://theiphonewiki.com/wiki/XCon*

7. *http://www.cultofmac.com/82097/ibooks-1-2-1-tries-to-run-jailbreak-code-to-detect-jailbroken-iphones/*

8. *http://theiphonewiki.com/wiki/Bypassing_Jailbreak_Detection*

9. *http://stackoverflow.com/questions/4165138/detect-udid-spoofing-on-the-iphone-at-runtime/*

Apple doesn't publicly disclose the techniques it uses to test applications for possible acceptance to the App Store, but it's clear that both binary analysis and dynamic testing are performed. This process has kept most blatant malware out of the App Store, at the cost of also barring any applications that Apple doesn't like the look of (including many types of communication apps, pornography, anything fun, and so on).

But despite Apple's efforts, it has been demonstrated that a moderately sophisticated attacker can get applications through App Store review while retaining the ability to download new code dynamically. There are a few different ways an attacker can approach this.

Bridging from WebKit

There are several approaches to accessing native iOS APIs via JavaScript, such as a user's location or use of media services, using a WebKit-based bridge. PhoneGap is a prominent example of such a package.[10] While these bridges can provide useful functionality and flexibility, using them also means that much application logic ends up in JavaScript and isn't necessarily shipped with the application to begin with. For example, a developer might implement a generic file-opening function that's accessible via JavaScript and avoid using it for anything evil during the review process. But later, that developer can alter the JavaScript served to the device and attempt to read data from areas on the device that they shouldn't be accessing.

I'll discuss the implementation of and some of the issues with JavaScript/native code bridges in Chapter 9.

Dynamic Patching

Normally, applications are prevented from running any native code that hasn't been cryptographically signed by Apple-issued keys. If a bug or misfeature in Apple's signature-checking logic is found, it can potentially allow for the downloading and execution of native code. A notable example of this in the wild was Charlie Miller's exploitation of a feature that allowed programs to allocate memory regions without NX protection (that is, memory regions that are readable, writable, and executable) and that do not require code to be signed.[11] This mechanism was put in place by Apple to allow Safari's JIT compiler to function,[12] but a bug in the implementation let third-party applications perform the same trick.

---

10. *http://phonegap.com/*

11. *http://arstechnica.com/apple/2011/11/safari-charlie-discovers-security-flaw-in-ios-gets-booted -from-dev-program/*

12. *http://reverse.put.as/wp-content/uploads/2011/06/syscan11_breaking_ios_code_signing.pdf*

This meant that native code could be downloaded and executed without needing to be signed at all. Miller demonstrated this by submitting an application, called *InstaStock*, to the App Store which purported to be a stock ticker checking program. At the time of app review, the app did nothing malicious or unusual; however, after the review process was complete, Miller was able to instruct the program to download new, unsigned code and execute that code without problem. This issue is now resolved, but it does give you an idea of the things that can slip through the cracks of review.

Intentionally Vulnerable Code

An interesting approach to bypassing App Store review is to intentionally make your app vulnerable to remote exploits. *Jekyll*[13] was a proof-of-concept application developed at Georgia Tech that intentionally introduced a buffer overflow in the core application. Malicious code was included in the app itself so that the code would be signed but was never called by the application. After approval, the researchers were able to use a buffer overflow exploit to change the control flow of the application to include malicious code, allowing it to use private Apple frameworks to interact with Bluetooth, SMS, and more.

Embedded Interpreters

While Apple's policy on this practice has shifted over the years, many products (primarily games) use an embedded Lua interpreter to perform much of the internal logic. Malicious behavior using an embedded interpreter has not yet been reported in the wild, but a crafty application using a similar interpreter could download code dynamically and execute it from memory, though not during the review process, of course. This would add new and malicious (or helpful, if you're so inclined) functionality.

Closing Thoughts

Ultimately, what protections does application review provide? Well, it does weed out less sophisticated malware. But you can assume with some certainty that malicious applications will indeed slip through from time to time. Keep that in mind and code your applications defensively; you definitely *cannot* assume other applications on the OS are benign.

---

13. *http://www.cc.gatech.edu/~klu38/publications/security13.pdf*

2

OBJECTIVE-C FOR THE LAZY

Objective-C has been met with both derision and adulation during its illustrious career. Brought to popularity by NeXTStep and inspired by the design of Smalltalk, Objective-C is a superset of C. Its most notable characteristics are the use of infix notation and absurdly long class names. People tend to either love it or hate it. People who hate it are wrong.

In this chapter, I'll go over the basics of Objective-C, assuming that you're already familiar with programming in some language or another. Know, however, that Cocoa and Objective-C are constantly changing. I can't cover all of their finer details adequately in a single chapter, but I do include some hints here to help nondevelopers get their bearings when examining Objective-C code. If you're starting from very little programming knowledge, you may wish to check out a book like Knaster, Malik, and Dalrymple's *Learn Objective-C on the Mac: For OS X and iOS* (Apress, 2012) before you dig in.

As much as I'd like to stick with the most modern coding patterns of Objective-C, if you're auditing existing code, you may come across plenty of crusty, reused code from the early days of iOS. So just in case, I'll go over both historical Objective-C constructs and the newly sanctioned versions.

Key iOS Programming Terminology

There are a few terms you'll want to be familiar with to understand where Apple's various APIs come from. *Cocoa* is the general term for the frameworks and APIs that are used in Objective-C GUI programming. *Cocoa Touch* is a superset of Cocoa, with some added mobile-related APIs such as dealing with gestures and mobile GUI elements. *Foundation* classes are Objective-C classes that make up much of what we call the Cocoa API. *Core Foundation* is a lower-level C-based library upon which many Foundation classes are based, usually prefixed with CF instead of NS.

Passing Messages

The first key to grokking Objective-C is understanding that the language is designed around the concept of *message passing*, rather than *calling*. It's useful (for me, anyway) to think of Objective-C as a language where objects sit around shouting at each other in a crowded room, rather than a language where hierarchical directors give orders to their subordinates. This analogy especially makes sense in the context of delegates, which I'll get to shortly.

At its most basic, sending Objective-C messages looks like this:

```
[Object doThisThingWithValue:myValue];
```

That's like saying, "Hey there, Object! Please do this thing using a value of myValue." When passing in multiple parameters, the nature of the first one is conventionally indicated by the message name. Any subsequent parameters must be both defined as part of the class and specifically named when called, as in this example:

```
if (pantsColor == @"Black") {

    [NSHouseCat sleepOnPerson:person
                    withRegion:[person lap]
                  andShedding:YES
                      retries:INT_MAX];
}
```

In this simplified simulation of catnapping under certain conditions, sleepOnPerson specifies a place to sleep (person), and withRegion specifies the region of the person to sleep on by sending person a message returning that person's lap. The andShedding parameter accepts a Boolean, and retries specifies the number of times this action will be attempted—in this case, up to the maximum value of an integer on a platform, which will vary depending on whether you have a 64-bit cat.

If you've been writing Objective-C for a while, you may notice that the formatting of this code looks different than what you're used to. That's because this is an arcane method of formatting Objective-C code, known as "the correct way," with vertically aligned colons between argument names and values. This keeps the pairings between parameter names and values visually obvious.

Dissecting an Objective-C Program

The two main parts of an Objective-C program are the *interface* and the *implementation*, stored in *.h* and *.m* files, respectively. (These are roughly analogous in purpose to *.h* and *.cpp* files in C++.) The former defines all of the classes and methods, while the latter defines the actual meat and logic of your program.

Declaring an Interface

Interfaces contain three main components: instance variables (or *ivars*), class methods, and instance methods. Listing 2-1 is the classic (that is, deprecated) Objective-C 1.0 way to declare your interfaces.

```
   @interface Classname : NSParentClass {
❶      NSSomeType aThing;
       int anotherThing;
   }
❷  + (type)classMethod:(vartype)myVariable;
❸  - (type)instanceMethod:(vartype)myVariable;
   @end
```

Listing 2-1: Declaring an interface, archaic version

Inside the main `@interface` block at ❶, instance variables are declared with a class (like `NSSomeType`) or a type (like `int`), followed by their name. In Objective-C, a + denotes the declaration of a class method ❷, while a - indicates an instance method ❸. As with C, the return type of a method is specified in parentheses at the beginning of the definition.

Of course, the modern way of declaring interfaces in Objective-C is a little different. Listing 2-2 shows an example.

```
❶  @interface Kitty : NSObject {
       @private NSString *name;
       @private NSURL *homepage;
       @public NSString *color;
   }

   @property NSString *name;
   @property NSURL *homepage;
❷  @property(readonly) NSString *color;
```

```
+ (type)classMethod:(vartype)myVariable;
- (type)instanceMethod:(vartype)myVariable;
```

Listing 2-2: Declaring an interface, modern version

This new class, called `Kitty`, inherits from `NSObject` ❶. `Kitty` has three instance variables of different accessibility types, and three properties are declared to match those instance variables. Notice that `color` is declared `readonly` ❷; that's because a `Kitty` object's color should never change. This means when the property is synthesized, only a getter method will be created, instead of both a getter and a setter. `Kitty` also has a pair of methods: one class method and one instance method.

You may have noticed that the example interface declaration used the `@private` and `@public` keywords when declaring instance variables. Similar to other languages, these keywords define whether ivars will be accessible from within only the class that declared it (`@private`), accessible from within the declaring class and any subclasses (`@protected`), or accessible by any class (`@public`). The default behavior of ivars is `@protected`.

NOTE *Newcomers to the language often want to know whether there is an equivalent to private methods. Strictly speaking, there isn't a concept of private methods in Objective-C. However, you can have the functional equivalent by declaring your methods only in the `@implementation` block instead of declaring them in both the `@interface` and the `@implementation`.*

Inside an Implementation File

Just like *.c* or *.cpp* files, Objective-C implementation files contain the meat of an Objective-C application. By convention, Objective-C files use *.m* files, while Objective-C++ files (which mix C++ and Objective-C code) are stored in *.mm* files. Listing 2-3 breaks down the implementation file for the `Kitty` interface in Listing 2-2.

```
@implementation Kitty
❶ @synthesize name;
@synthesize color;
@synthesize homepage;

+ (type)classMethod:(vartype)myVariable {
    // method logic
}

- (type)instanceMethod:(vartype)myVariable {
    // method logic
}
@end
```

```
Kitty *myKitty = [[Kitty alloc] init];
```

❷ `[myKitty setName:@"Ken"];`
❸ `myKitty.homepage = [[NSURL alloc] initWithString:@"http://me.ow"];`

Listing 2-3: A sample implementation

The `@synthesize` statements at ❶ create the setter and getter methods for the properties. Later, these getter and setter methods can be used either with Objective-C's traditional infix notation ❷, where methods of the format *propertyName* and *setPropertyName* (like `name` and `setName`, respectively) get and set values, or with dot notation ❸, where properties like `homepage` are set or read using the *.property* format, as they might be in other languages.

NOTE   *Be careful with dot notation, or just don't use it. Dot notation makes it hard to know whether you're dealing with an object or a C struct, and you can actually call any method with it—not only getters and setters. Dot notation is also just visually inconsistent. Long story short, in this book I'll avoid dot notation in the name of consistency and ideological purity. But despite my best efforts, you'll likely encounter it in the real world anyway.*

Technically, you don't need to synthesize properties that are declared in the interface file with `@property`, like `name`, `color`, and `homepage` in Listing 2-3; the compiler in recent versions of Xcode synthesizes these properties on its own. But you may want to manually declare them anyway for clarity or when you want to change the name of the instance variable to differentiate it from the property name. Here's how manually synthesizing a property works:

```
@synthesize name = thisCatName;
```

Here, the property `name` is backed by the instance variable `thisCatName` because it was manually synthesized. However, the default behavior with automatic property synthesis is analogous to this:

```
@synthesize name = _name;
```

This default behavior prevents developers from accidentally meddling with the instance variables directly, instead of using setters and getters, which can cause confusion. For example, if you set an ivar directly, you'll be bypassing any logic in your setter/getter methods. Automatic synthesis is probably the best way to do things, but you'll be seeing manual synthesis in code for a long time to come, so it's best to be aware of it.

Specifying Callbacks with Blocks

One thing that's becoming increasingly popular in Objective-C code is the use of *blocks*, which are often used in Cocoa as a way to specify a callback. For example, here's how you'd use the dataTaskWithRequest method of the NSURLSessionDataTask class:

```
NSURLSession *session = [NSURLSession sessionWithConfiguration:configuration
                                       delegate:self
                                       delegateQueue:nil];

NSURLSessionDataTask *task = [session dataTaskWithRequest:request
                                      completionHandler:
❶      ^(NSData *data, NSURLResponse *response, NSError *error) {
            NSLog(@"Error: %@ %@", error, [error userInfo]);
        }];
```

The ^ at ❶ is declaring a block that will be executed once the request is complete. Note that no name is specified for this function because it won't be called from anywhere other than this bit of code. A block declaration just needs to specify the parameters that the closure will take. From there, the rest of the block is just like a normal function. You can use blocks for tons of other things as well, but to start with, it's probably sufficient to have a basic understanding of what they are: things that begin with ^ and do stuff.

How Objective-C Manages Memory

Unlike some other languages, Objective-C does not have any garbage collection. Historically, Objective-C has used a *reference counting model*, using the retain and release directives to indicate when an object needs to be freed, to avoid memory leaks. When you retain an object, you increase the *reference count*—that is, the number of things that want that object to be available to them. When a piece of code no longer needs an object, it sends it a release method. When the reference count reaches zero, the object is deallocated, as in this example:

```
❶  NSFish *fish = [[NSFish alloc] init];
   NSString *fishName = [fish name];
❷  [fish release];
```

Assume that before this code runs, the reference count is 0. After ❶, the reference count is 1. At ❷, the release method is called to say that the fish object is no longer needed (the application just needs the fish object's name property), and when fish is released, the reference count should be 0 again.

The [[Classname alloc] init] can also be shortened to [Classname new], but the new method isn't favored by the Objective-C community because it's less explicit and is inconsistent with methods of object creation other than init. For example, you can initialize NSString objects with [[NSString alloc] initWithString:@"My string"] , but there's no equivalent new syntax, so your code would end up having a mix of both methods. Not everyone is averse to new, and it's really a matter of taste, so you're likely to see it both ways. But in this book, I'll favor the traditional approach.

Regardless of which allocation syntax you prefer, the problem with a manual retain/release is that it introduced the possibility of errors: programmers could accidentally release objects that had already been deallocated (causing a crash) or forget to release objects (causing a memory leak). Apple attempted to simplify the situation with automatic reference counting.

Automatic Reference Counting

Automatic reference counting (ARC) is the modern method of Objective-C memory management. It removes the need for manually tracking reference counts by automatically incrementing and decrementing the retain count where appropriate.[1] Essentially, it inserts retain and release methods for you. ARC introduces a few new concepts, listed here:

- *Weak* and *strong* references assist in preventing cyclical references (referred to as *strong reference cycles*), where a parent object and child object both have ownership over each other and never get deallocated.

- Object ownership between Core Foundation objects and Cocoa objects can be bridged. Bridging tells the compiler that Core Foundation objects that are cast to Cocoa objects are to be managed by ARC, by using the __bridge family of keywords.

- @autoreleasepool replaces the previously used NSAutoReleasePool mechanism.

In modern Cocoa applications with ARC, the details of memory management are unlikely to come into play in a security context. Previously exploitable conditions such as double-releases are no longer a problem, and memory-management-related crashes are rare. It's still worth noting that there are other ways to cause memory management problems because CFRetain and CFRelease still exist for Core Foundation objects and C malloc and free can still be used. I'll discuss potential memory management issues using these lower-level APIs in Chapter 11.

---

1. *http://developer.apple.com/library/mac/#releasenotes/ObjectiveC/RN-TransitioningToARC/ Introduction/Introduction.html*

Delegates and Protocols

Remember how objects "shout at each other in a crowded room" to pass messages? *Delegation* is a feature that illustrates Objective-C's message-passing architecture particularly well. Delegates are objects that can receive messages sent during program execution and respond with instructions that influence the program's behavior.

To be a delegate, an object must implement some or all methods defined by a *delegate protocol*, which is an agreed-upon method of communication between a delegator and a delegate. You can declare your own protocols, but most commonly you'll be using established protocols in the core APIs.

The delegates you'll write will typically respond to one of three fundamental message types: *should*, *will*, and *did*. Invoke these messages whenever an event is about to happen and then let your delegates direct your program to the correct course of action.

Should Messages

Objects pass *should* messages to request input from any available delegates on whether letting an event happen is a good idea. Think of this as the final call for objections. For example, when a `shouldSaveApplicationState` message is invoked, if you've implemented a delegate to handle this message, your delegate can perform some logic and say something like, "No, actually, we shouldn't save the application state because the user has checked a checkbox saying not to." These messages generally expect a Boolean as a response.

Will Messages

A *will* message gives you the chance to perform some action before an event occurs—and, sometimes, to put the brakes on before it does. This message type is more like saying, "Hey guys! Just an FYI, but I'm going to go do this thing, unless you need to do something else first. I'm pretty committed to the idea, but if it's a total deal-breaker, let me know and I can stop." An example would be the `applicationWillTerminate` message.

Did Messages

A *did* message indicates that something has been decided for sure and an event is going to happen whether you like it or not. It also indicates that if any delegates want to do some stuff as a result, they should go right ahead. An example would be `applicationDidEnterBackground`. In this case, did isn't really an indication that the application *has* entered the background, but it's a reflection of the decision being definitively made.

Declaring and Conforming to Protocols

To declare that your class conforms to a protocol, specify that protocol in your @interface declaration within angle brackets. To see this in action, look at Listing 2-4, which shows an example @interface declaration that uses the NSCoding protocol. This protocol simply specifies that a class implements two methods used to encode or decode data: encodeWithCoder to encode data and initWithCoder to decode data.

```
❶ @interface Kitty : NSObject <NSCoding> {
       @private NSString *name;
       @private NSURL *homepage;
       @public NSString *color;
   }

   @implementation Kitty

❷ - (id)initWithCoder:(NSCoder *)decoder {
       self = [super init];
       if (!self) {
           return nil;
       }

       [self setName:[decoder decodeObjectForKey:@"name"]];
       [self setHomepage:[decoder decodeObjectForKey:@"homepage"]];
       [self setColor:[decoder decodeObjectForKey:@"color"]];

       return self;
   }

❸ - (void)encodeWithCoder:(NSCoder *)encoder {
       [encoder encodeObject:[self name] forKey:@"name"];
       [encoder encodeObject:[self author] forKey:@"homepage"];
       [encoder encodeObject:[self pageCount] forKey:@"color"];
   }
```

Listing 2-4: Declaring and implementing conformance to the NSCoding protocol

The declaration at ❶ specifies that the Kitty class will be conforming to the NSCoding protocol.[2] When a class declares a protocol, however, it must also conform to it, which is why Kitty implements the required initWithCoder ❷ and encodeWithCoder ❸ methods. These particular methods are used to serialize and deserialize objects.

---

2. *https://developer.apple.com/library/mac/documentation/Cocoa/Reference/Foundation/Protocols/NSCoding_Protocol/Reference/Reference.html*

If none of the built-in message protocols do what you need, then you can also define your own protocols. Check out the declaration of the NSCoding protocol in Apple's Framework header files (Listing 2-5) to see what a protocol definition looks like.

```
@protocol NSCoding

- (void)encodeWithCoder:(NSCoder *)aCoder;
- (id)initWithCoder:(NSCoder *)aDecoder;

@end
```

Listing 2-5: The declaration of the NSCoding protocol, from Frameworks/NSCoding.h

Notice that the NSCoding definition contains two methods that any class conforming to this protocol must implement: encodeWithCoder and initWithCoder. When you define a protocol, you must specify those methods yourself.

The Dangers of Categories

Objective-C's *category* mechanism allows you to implement new methods on existing classes at runtime, without having to recompile those classes. Categories can add or replace methods in the affected class, and they can appear anywhere in the codebase. It's an easy way to quickly change the behavior of a class without having to reimplement it.

Unfortunately, using categories is also an easy way to make egregious security mistakes. Because they can affect your classes from anywhere within the codebase—even if they appear only in third-party code—critical functionality, such as TLS endpoint validation, can be completely overridden by a random third-party library or a careless developer. I've seen this happen in important iOS products before: after carefully verifying that TLS/SSL works correctly in their application, developers include a third-party library that overrides that behavior, messing up their own properly designed code.

You can usually spot categories by noting @implementation directives that purport to implement classes already present in Cocoa Touch. If a developer was actually creating a category there, then the name of the category would follow the @implementation directive in parentheses (see Listing 2-6).

```
@implementation NSURL (CategoryName)

- (BOOL) isPurple; {
    if ([self isColor:@"purple"])
        return YES;
```

```
        else
            return NO;
    }
    @end
```

Listing 2-6: Implementing a category method

You can also use categories to override *existing* class methods, which is a potentially useful but particularly dangerous approach. This can cause security mechanisms to be disabled (such as the aforementioned TLS validation) and can also result in unpredictable behavior. Quoth Apple:

> If the name of a method declared in a category is the same as a method in the original class, or a method in another category on the same class (or even a superclass), the behavior is undefined as to which method implementation is used at runtime.

In other words, multiple categories can define or overwrite the same method, but only one will "win" and be called. Note that some Framework methods may themselves be implemented via a category—if you attempt to override them, your category *might* be called, but it might not.

A category may also accidentally override the functionality of subclasses, even when you only meant for it to add a new method. For example, if you were to define an isPurple method on NSObject, all subclasses of NSObject (which is to say, all Cocoa objects) would inherit this method. Any other class that defined a method with the same name might or might not have its method implementation clobbered. So, yes, categories are handy, but use them sparingly; they can cause serious confusion as well as security side effects.

Method Swizzling

Method swizzling is a mechanism by which you can replace the implementation of a class or instance method that you don't own (that is, a method provided by the Cocoa API itself). Method swizzling can be functionally similar to categories or subclassing, but it gives you some extra power and flexibility by actually swapping the implementation of a method with a totally new implementation, rather than extending it. Developers typically use this technique to augment functionality of a method that's used by many different subclasses so they don't have to duplicate code.

The code in Listing 2-7 uses method swizzling to add a logging statement to any call of setHidden. This will affect any subclass of UIView, including UITextView, UITextField, and so forth.

```
#import <objc/runtime.h>

@implementation UIView(Loghiding)
```
❶
```
- (BOOL)swizzled_setHidden {
    NSLog(@"We're calling setHidden now!");
```
❷
```
    BOOL result = [self swizzled_setHidden];

    return result;
}
```
❸
```
+ (void)load {
    Method original_setHidden;
    Method swizzled_setHidden;

    original_setHidden = class_getInstanceMethod(self, @selector(setHidden));
    swizzled_setHidden = class_getInstanceMethod(self, @selector(swizzled_
      setHidden));
```
❹
```
    method_exchangeImplementations(original_setHidden, swizzled_setHidden);
}
```
```
@end
```

Listing 2-7: Exchanging the implementation of an existing method and a replacement method

At ❶, a wrapper method is defined that simply spits out an SLog that the setHidden method is being called. But at ❷, the swizzle_SetHidden method appears to be calling itself. That's because it's considered a best practice to call the original method after performing any added functionality, to prevent unpredictable behavior like failing to return the type of value the caller would expect. When you call swizzled_setHidden from within itself, it actually calls the *original* method because the original method and the replacement method have already been swapped.

The actual swapping is done in the load class method ❸, which is called by the Objective-C runtime when loading the class for the first time. After the references to the original and swizzled methods are obtained, the method_exchangeImplementations method is called at ❹, which, as the name implies, swaps the original implementation for the swizzled one.

There are a few different strategies for implementing method swizzling, but most of them carry some risk since you're mucking around with core functionality.

If you or a loved one want to implement method swizzling, you may want to consider using a fairly well-tested wrapper package, such as JRSwizzle.[3] Apple may reject applications that appear to use method swizzling in a dangerous way.

Closing Thoughts

Overall, Objective-C and the Cocoa API are nicely high-level and prevent a number of classic security issues in C. While there are still several ways to mess up memory management and object manipulation, most of these methods result in a denial of service at worst in modern code. If you're a developer, rely on Cocoa as much as possible, rather than patching in C or C++ code.

Objective-C does, however, contain some mechanisms, such as categories or swizzling, that can cause unexpected behavior, and these mechanisms can affect your codebase widely. Be sure to investigate these techniques when you see them during an app assessment because they can potentially cause some serious security misbehavior.

---

3. *https://github.com/rentzsch/jrswizzle/*

3

IOS APPLICATION ANATOMY

To understand some of the problems iOS applications face, it's useful to get an idea of how different types of data are stored and manipulated within an application's private directory, where all of its configuration, assets, binaries, and documents are stored. This is where you can discover all manner of information leakage, as well as dig in to the guts of the program that you're examining.

The quickest way find out what data your application stores locally on an iOS device is to check out *~Library/Developer/CoreSimulator/Devices*. Starting with Xcode 6, each combination of device type and OS version you've ever deployed into the Simulator application is assigned a UUID. Your particular application's data will be stored in two places under this directory. Your application binary and assets, including *.nib* user interface files and graphic files included with the application, are in *<device ID>/data/Containers/Bundle/Application/<app bundle id>*. The more dynamic data that your application stores is in *~<device ID>/ data/Containers/Data/Application/<app bundle id>*. Systemwide data such as global configurations will be stored in the remainder of the *<device ID>* directory.

Exploring this directory structure, which is sketched out in simplified form in Figure 3-1, also reveals which types of data are handled by OS services rather than your application.

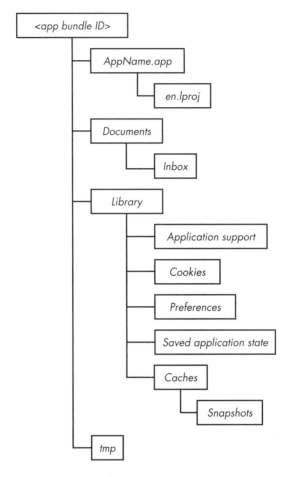

Figure 3-1: Layout of an application directory

If you're on a jailbroken device, you can use SSH to connect to the device and explore the directory structure; I'll talk about jailbreaking and connecting to test devices in Chapter 6. Whether or not your device is jailbroken, you can use a tool such as iExplorer[1] to examine the directory structure of your installed applications, as shown in Figure 3-2.

In the rest of this chapter, I'll cover some of the common directories and data stores used by iOS applications, as well as how to interact with them programmatically and what data can leak from them.

---

1. *http://www.macroplant.com/iexplorer/*

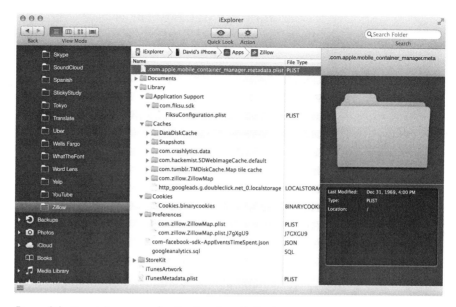

Figure 3-2: Examining an application bundle with iExplorer

Dealing with plist Files

Before you start examining the directory structure, you need to know how to read some of the stuff you'll find there. iOS stores app configuration data inside *property list (plist)* files, which hold this information in Core Foundation data types such as CFArray, CFString, and so forth. From a security standpoint, you want to examine plists for things that shouldn't be stored in plaintext, such as credentials, and then potentially manipulate them to change the application's behavior. For instance, you could enable a paid feature that's disabled.

There are two types of plist formats: binary and XML. As you can see in the following example, the XML format is easily readable by humans.

```
<?xml version="1.0" encoding="UTF-8"?>
<!DOCTYPE plist PUBLIC "-//Apple//DTD PLIST 1.0//EN" "http://www.apple.com/DTDs/
    PropertyList-1.0.dtd">
<plist version="1.0">
<dict>
<plist version="1.0">
<dict>
  <key>CFBundleDevelopmentRegion</key>
  <string>en</string>
  <key>CFBundleExecutable</key>
  <string>Test</string>
  <key>CFBundleIdentifier</key>
  <string>com.dthiel.Test</string>
  <key>CFBundleInfoDictionaryVersion</key>
```

```
<string>6.0</string>
<key>CFBundleName</key>
<string>Test</string>
<key>CFBundlePackageType</key>
<string>APPL</string>
<key>CFBundleShortVersionString</key>
<string>1.0</string>
<key>CFBundleSignature</key>
<string>????</string>
<key>CFBundleSupportedPlatforms</key>
<array>
  <string>iPhoneSimulator</string>
</array>
--snip--
```

This is simply a dictionary containing hierarchical keys and values, which provides information about the app—the platforms it can run on, the code signature, and so forth (the signature is not present here because the app is deployed in the Simulator application).

But when examining files from the command line or working with plists programmatically, you'll frequently encounter plists in binary format, which is not particularly human readable (or writable). You can convert these plists to XML using the plutil(1) command.

```
$ plutil -convert xml1 Info.plist -o -
$ plutil -convert xml1 Info.plist -o Info-xml.plist
$ plutil -convert binary1 Info-xml.plist -o Info-bin.plist
```

The first command converts a binary plist to XML and outputs it to stdout, where you can pipe it to less(1) or similar commands. You can also output directly to a file with -o *filename*, as in the second command. In the third command, the binary1 conversion type turns an XML-formatted plist to binary; but since the formats are interchangeable, you shouldn't really need to do this.

To make reading and editing plists more seamless, you can also configure your text editor to automatically convert plist files so that if you need to read or write to one, you can do so smoothly from a familiar environment. For example, if you happen to use Vim, you might add a configuration like this to your *.vimrc* file:

```
" Some quick bindings to edit binary plists
command -bar PlistXML :set binary | :1,$!plutil -convert xml1 /dev/stdin -o -
command -bar Plistbin :1,$!plutil -convert binary1 /dev/stdin -o -

fun ReadPlist()
    if getline("'[") =~ "^bplist"
        :PlistXML
```

```
        set filetype=xml
    endif
endfunction

augroup misc
    au BufWinEnter *.plist, call ReadPlist()
augroup end
```

This configuration will use the `:PlistXML` command to automatically convert any binary plist that you edit to XML format, allowing you to make changes in a human-readable format. Before actually writing those changes to the file, the configuration will convert the file to binary again using the `:Plistbin` command. Note that the file will still be successfully consumed by applications regardless of whether it is in binary or XML format.

You can view plists of either format within Xcode, as in Figure 3-3. The advantage of using Xcode is that you'll have some additional help and drop-down menus that show you what potential values you might be able to use for the various keys. It's good to know how to work with plists from the command line, though, because this lets you directly interact with them via SSH sessions to jailbroken devices.

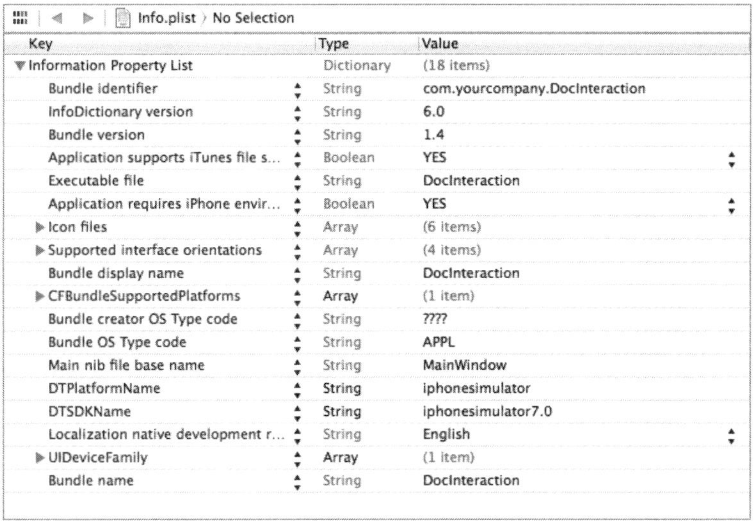

Key	Type	Value
▼ Information Property List	Dictionary	(18 items)
Bundle identifier	String	com.yourcompany.DocInteraction
InfoDictionary version	String	6.0
Bundle version	String	1.4
Application supports iTunes file s...	Boolean	YES
Executable file	String	DocInteraction
Application requires iPhone envir...	Boolean	YES
▶ Icon files	Array	(6 items)
▶ Supported interface orientations	Array	(4 items)
Bundle display name	String	DocInteraction
▶ CFBundleSupportedPlatforms	Array	(1 item)
Bundle creator OS Type code	String	????
Bundle OS Type code	String	APPL
Main nib file base name	String	MainWindow
DTPlatformName	String	iphonesimulator
DTSDKName	String	iphonesimulator7.0
Localization native development r...	String	English
▶ UIDeviceFamily	Array	(1 item)
Bundle name	String	DocInteraction

Figure 3-3: Viewing a plist within Xcode

See the man pages `plist(5)` and `plutil(1)` for more information about viewing and editing plists. If you're working on a jailbroken device, you can use the `plutil` command included with Erica Sadun's Erica Utilities[2] (available in Cydia) to work with plists locally.

_____

2. Erica Utilities has a number of other useful tools for working with jailbroken devices; you can check out the list at *http://ericasadun.com/ftp/EricaUtilities/*.

Device Directories

Starting with iOS 8, Simulator platforms such as iPhone, iPad, and their variations are stored in directories named with unique identifiers. These identifiers correspond with the type of device you choose when launching the Simulator from Xcode, in combination with the requested OS version. Each of these directories has a plist file that describes the device. Here's an example:

```
<?xml version="1.0" encoding="UTF-8"?>
<!DOCTYPE plist PUBLIC "-//Apple//DTD PLIST 1.0//EN" "http://www.apple.com/DTDs/
    PropertyList-1.0.dtd">
<plist version="1.0">
<dict>
        <key>UDID</key>
        <string>DF15DA82-1B06-422F-860D-84DCB6165D3C</string>
        <key>deviceType</key>
        <string>com.apple.CoreSimulator.SimDeviceType.iPad-2</string>
        <key>name</key>
        <string>iPad 2</string>
        <key>runtime</key>
        <string>com.apple.CoreSimulator.SimRuntime.iOS-8-0</string>
        <key>state</key>
        <integer>3</integer>
</dict>
</plist>
```

In this plist file, it's not immediately obvious which directory is for which device. To figure that out, either you can look at the *.default_created.plist* file in the *Devices* directory, or you can just grep all of the *device.plist* files, as shown in Listing 3-1.

```
$ cd /Users/me/Library/Developer/CoreSimulator/Devices && ls
26E45178-F483-4CDD-A619-9C0780293DD4
78CAAF2B-4C54-4519-A888-0DB84A883723
A2CD467D-E110-4E38-A4D9-5C082618604A
AD45A031-2412-4E83-9613-8944F8BFCE42
676931A8-FDA5-4BDC-85CC-FB9E1B5368B6
989328FA-57FA-430C-A71E-BE0ACF278786
AA9B1492-ADFE-4375-98F1-7DB53FF1EC44
DF15DA82-1B06-422F-860D-84DCB6165D3C

$ for dir in `ls|grep -v default`
do
echo $dir
grep -C1 name $dir/device.plist |tail -1|sed -e 's/<\/*string>//g'
done
```

```
26E45178-F483-4CDD-A619-9C0780293DD4
        iPhone 5s
676931A8-FDA5-4BDC-85CC-FB9E1B5368B6
        iPhone 5
78CAAF2B-4C54-4519-A888-0DB84A883723
        iPad Air
989328FA-57FA-430C-A71E-BE0ACF278786
        iPhone 4s
A2CD467D-E110-4E38-A4D9-5C082618604A
        iPad Retina
AA9B1492-ADFE-4375-98F1-7DB53FF1EC44
        Resizable iPad
AD45A031-2412-4E83-9613-8944F8BFCE42
        Resizable iPhone
DF15DA82-1B06-422F-860D-84DCB6165D3C
        iPad 2
```

Listing 3-1: Grepping to determine which identifier maps to which model of iOS device

After entering the appropriate directory for the device you've been testing your application on, you'll see a *data* directory that contains all of the Simulator files, including those specific to your application. Your application data is split into three main directories under *data/Containers*: *Bundle*, *Data*, and *Shared*.

The Bundle Directory

The *Bundle* directory contains an *Applications* directory, which in turn contains a directory for each of the applications stored on the device, represented by that application's bundle ID. In each application's directory, the *.app* folder is where the application's core binary is stored, along with image assets, localization information, and the *Info.plist* file that contains the core configuration information for your application. *Info.plist* includes the bundle identifier and main executable, along with information about your application's UI and which device capabilities an application requires to be able to run.

On the filesystem, these plists are stored in either XML or binary format, with the latter being the default. You can retrieve the information in *Info.plist* programmatically by referencing dictionary attributes of [NSBundle mainBundle];[3] this is commonly used for loading styling or localization information.

---

3. *https://developer.apple.com/library/Mac/documentation/Cocoa/Reference/Foundation/Classes/ NSBundle_Class/Reference/Reference.html*

One thing that will potentially be of interest in the *Info.plist* file is the UIRequiredDeviceCapabilities entry, which looks something like this:

```
<key>UIRequiredDeviceCapabilities</key>
<dict>
        <key>armv7</key>
        <true/>
        <key>location-services</key>
        <true/>
        <key>sms</key>
        <true/>
</dict>
```

The UIRequiredDeviceCapabilities entry describes which system resources an app requires. While not an enforcement mechanism, this can give you some clues as to what type of activities the application will engage in.

The Data Directory

The primary area of interest in the *Data* directory is the *Applications* subdirectory. The *Data/Applications* directory contains the rest of the data an application uses to run: preferences, caches, cookies, and so on. This is also the primary location you'll want to inspect for most types of data leakage. Now, let's go over the various subdirectories and the types of data that they may end up holding.[4]

The Documents and Inbox Directories

The *Documents* directory is intended to store your nontransient application data, such as user-created content or local information allowing the app to run in offline mode. If UIFileSharingEnabled is set in your application's *Info.plist* file, files here will be accessible via iTunes.

Data files that other applications ask your app to open are stored in your application's *Documents/Inbox* directory. These will be invoked by the calling application using the UIDocumentInteractionController class.[5]

You can only read or delete files stored in the *Inbox* directory. These files come from another application that can't write to your app directory, so they're put there by a higher-privileged system process. You may want to consider deleting these files periodically or giving the user the option to delete them because it will not be apparent to the user what documents are stored here and whether they contain sensitive information.

---

4. Note that not all directories that can exist in this directory tree will exist for every application; some are created on the fly only when certain APIs are used by the app.

5. *http://developer.apple.com/library/ios/#documentation/FileManagement/Conceptual/DocumentInteraction_TopicsForIOS*

If you're writing an application with the goal of ensuring sensitive information doesn't remain on disk, copy documents out of the *Inbox* directory to a separate location where you can apply Data Protection and then remove those files from the *Inbox* directory.

It's also worth remembering that under certain circumstances, any file your application asks to open may persist on the disk *forever*. If you attempt to open a file type that your program isn't a handler for, then that file will be passed off to a third-party app, and who knows when the other app will delete it? It may get stored indefinitely. In other words, the cleanup of any file that you ask another app to open is beyond your control, even if you simply preview the contents using the Quick Look API. If having *Inbox* files kick around for a long time is problematic, consider giving your application the ability to view such data on its own (rather than relying on a helper) and then make sure to dispose of the files properly.

The Library Directory

The *Library* directory contains the majority of your application's files, including data cached by the application or by particular networking constructs. It will be backed up via iTunes and to iCloud, with the exception of the *Caches* directory.

The Application Support Directory

The *Application Support* directory is not for storing files created or received by the user but rather for storing additional data files that will be used by your application. Examples would be additional purchased downloadable content, configuration files, high scores, and so on—as the name implies, things that support the running and operation of the application. Either these files can be deployed when the application is first installed or they can be downloaded or created by your application later.

By default, iTunes backs up the data in this directory to your computer and to iCloud. However, if you have privacy or security concerns about this data being stored in Apple's cloud environment, you can explicitly disallow this by setting the `NSURLIsExcludedFromBackupKey` attribute on newly created files. I'll discuss preventing data from syncing to iCloud further in Chapter 10.

Note that Apple requires that applications back up only user data to iCloud (including documents they've created, configuration files, and so forth), never application data. Applications that allow application content, such as downloadable app content, to be backed up to iCloud can be rejected from the App Store.

The Caches and Snapshots Directories

The *Caches* directory is similar in function to a web browser's cache: it's intended for data that your application will keep around for performance reasons but not for data that is crucial for the application to function. As such, this directory won't be backed up by iTunes.

While Apple states that your application is responsible for managing the *Caches* directory, the OS does actually manipulate the directory's contents and that of its subfolder, *Snapshots.* Always consider the contents of the *Caches* directory to be transient, and expect it to disappear between program launches. iOS will cull these cache directories automatically if the system starts running low on space, though it won't do this while the application is running.

The *Caches* directory also sometimes stores web cache content in a subdirectory such as *Caches/com.mycompany.myapp.* This is one place where sensitive data can leak because iOS can cache information delivered over HTTPS for quite a long time. If the developer hasn't made special effort to prevent data from being cached or to expire cached data quickly, you can often find some goodies in here.

Finally, when an application is put into the background, the OS also automatically stores screenshots of the application in the *Snapshots* sub-directory, potentially leaving sensitive information on local storage. This is done for one reason: so that the OS can use the current screen state to create the "whooshing" animation that happens when you bring an application to the foreground. Unfortunately, a side effect I frequently see in iOS applications is that the disk stores images of people's Social Security numbers, user details, and so on. I'll discuss mitigation strategies for this (and many other caching problems) in Chapter 10.

The Cookies Directory

The *Cookies* directory stores cookies set by the URL loading system. When you make an NSURLRequest, any cookies will be set according to either the default system cookie policy or one that you've specified. Unlike on OS X, cookies on iOS are not shared between applications; each application will have its own cookie store in this directory.

The Preferences Directory

iOS stores application preferences in the *Preferences* directory, but it doesn't allow applications to write directly to the files there. Instead, files in this directory are created, read, and manipulated by either the NSUserDefaults or CFPreferences API.

These APIs store application preference files in plaintext; therefore, you most definitely should *not* use them to store sensitive user information or credentials. When examining an application to see what information it's storing locally, be sure to examine the plist files in the *Preferences* directory. You'll sometimes find usernames and passwords, API access keys, or security controls that are not meant to be changed by users.

The Saved Application State Directory

Users expect apps to remember what they enter into text fields, which settings they've enabled, and so on. If a user switches to another application and then restores the original application at a later time, the application may have actually been killed by the operating system during the interval. To make it so that the UI remains consistent between program launches, recent versions of iOS store object state information in the *Saved Application State* directory by the State Preservation API.[6] Developers can tag specific parts of their UI to be included in State Preservation.

If you're not careful about what you store as part of the application state, this is one place you can wind up with data leaks. I'll discuss how to avoid those in depth in Chapter 10.

The tmp Directory

As you might surmise, *tmp* is where you store transient files. Like the *Caches* directory, the files contained in this directory may be automatically removed by the OS while your application isn't running. The usage of this directory is fairly similar to that of the *Caches* directory; the difference is that *Caches* is meant to be used for files that might need to be retrieved again or re-created. For example, if you download certain application data from a remote server and want to keep it around for performance reasons, you'd store that in *Caches* and redownload it if it disappears. On the other hand, *tmp* is for strictly temporary files generated by the application—in other words, files that you won't miss if they're deleted before you can revisit them. Also, like the *Caches* directory, *tmp* is not backed up to iTunes or iCloud.

The Shared Directory

The *Shared* directory is a bit of a special case. It's for applications that share a particular app group (introduced in iOS 8 to support extensions), such as those that modify the behavior of the Today screen or keyboard. Apple requires all extensions to have a container application, which receives its own app ID. The *Shared* directory is the way that the extension and its containing app share data. For example, apps can access databases of shared user defaults by specifying a suite name during initialization of NSUserDefaults, like this:

```
[[NSUserDefaults alloc] initWithSuiteName:@"com.myorg.mysharedstuff"];
```

While the *Shared* directory isn't commonly used at the time of writing, it's prudent to check this directory when looking for any sensitive information potentially stored in preferences or other private data.

---

6. *http://developer.apple.com/library/ios/documentation/iPhone/Conceptual/iPhoneOSProgrammingGuide/iPhoneAppProgrammingGuide.pdf* (page 69)

Closing Thoughts

With a basic understanding of the iOS security model, the Cocoa API, and how iOS applications are laid out, you're now ready to move on to the fun stuff: tearing apart applications and finding their flaws. In Part II, I'll show you how to build your testing platform, debug and profile applications, and deal with testing third-party apps for which source code is available.

PART II

SECURITY TESTING

4

BUILDING YOUR TEST PLATFORM

In this chapter, I'll outline the tools you need to review your code and test your iOS applications, and I'll show you how to build a robust and useful test platform. That test platform will include a properly set up Xcode instance, an interactive network proxy, reverse engineering tools, and tools to bypass iOS platform security checks.

I'll also cover the settings you need to change in Xcode projects to make bugs easier to identify and fix. You'll then learn to leverage Xcode's static analyzer and compiler options to produce well-protected binaries and perform more in-depth bug detection.

Taking Off the Training Wheels

A number of behaviors in a default OS X install prevent you from really digging in to the system internals. To get your OS to stop hiding the things you need, enter the following commands at a Terminal prompt:

```
$ defaults write com.apple.Finder AppleShowAllFiles TRUE
$ defaults write com.apple.Finder ShowPathbar -bool true
$ defaults write com.apple.Finder _FXShowPosixPathInTitle -bool true
```

```
$ defaults write NSGlobalDomain AppleShowAllExtensions -bool true
$ chflags nohidden ~/Library/
```

These settings let you see all the files in the Finder, even ones that are hidden from view because they have a dot in front of their name. In addition, these changes will display more path information and file extensions, and most importantly, they allow you to see your user-specific *Library*, which is where the iOS Simulator will store all of its data.

The chflags command removes a level of obfuscation that Apple has put on directories that it considers too complicated for you, such as */tmp* or */usr.* I'm using the command here to show the contents of the iOS Simulator directories without having to use the command line every time.

One other thing: consider adding *$SIMPATH* to the Finder's sidebar for easy access. It's convenient to use *$SIMPATH* to examine the iOS Simulator's filesystem, but you can't get to it in the Finder by default. To make this change, browse to the following directory in the Terminal:

```
$ cd ~/Library/Application\ Support
$ open .
```

Then, in the Finder window that opens, drag the iPhone Simulator directory to the sidebar. Once you're riding without training wheels, it's time to choose your testing device.

Suggested Testing Devices

My favorite test device is the Wi-Fi only iPad because it's inexpensive and easy to jailbreak, which allows for testing iPad, iPhone, and iPod Touch applications. Its lack of cellular-based networking isn't much of a hindrance, given that you'll want to intercept network traffic most of the time anyway.

But this configuration does have some minor limitations. Most significantly, the iPad doesn't have GPS or SMS, and it obviously doesn't make phone calls. So it's not a bad idea to have an actual iPhone of some kind available.

I prefer to have at least two iPads handy for iOS testing: one jailbroken and one stock. The stock device allows for testing in a legitimate, realistic end-user environment, and it has all platform security mechanisms still intact. It can also register properly for push notifications, which has proven problematic for jailbroken devices in the past. The jailbroken device allows you to more closely inspect the filesystem layout and more detailed workings of iOS; it also facilitates black-box testing that wouldn't be feasible using a stock device alone.

Testing with a Device vs. Using a Simulator

Unlike some other mobile operating systems, iOS development uses a *simulator* rather than an emulator. This means there's no full emulation of the iOS device because that would require a virtualized ARM environment. Instead, the simulators that Apple distributes with Xcode are compiled for the x64 architecture, and they run natively on your development machine, which makes the process significantly faster and easier. (Try to boot the Android emulator inside a virtual machine, and you'll appreciate this feature.)

On the flip side, some things simply don't work the same way in the iOS Simulator as they do on the device. The differences are as follows:

Case-sensitivity   Unless you've intentionally changed this behavior, OS X systems operate with case-insensitive HFS+ filesystems, while iOS uses the case-sensitive variant. This should rarely be relevant to security but can cause interoperability issues when modifying programs.

Libraries   In some cases, iOS Simulator binaries link to OS X frameworks that may behave differently than those on iOS. This can result in slightly different behavior.

Memory and performance   Since applications run natively in the iOS Simulator, they'll be taking full advantage of your development machine's resources. When gauging the impact of things such as PBKDF2 rounds (see Chapter 13), you'll want to compensate for this or test on a real device.

Camera   As of now, the iOS Simulator does not use your development machine's camera. This is rarely a huge issue, but some applications do contain functionality such as "Take a picture of my check stub or receipt," where the handling of this photo data can be crucial.

SMS and cellular   You can't test interaction with phone calls or SMS integration with the iOS Simulator, though you can technically simulate some aspects, such as toggling the "in-call" status bar.

Unlike in older versions of iOS, modern versions of the iOS Simulator do in fact simulate the Keychain API, meaning you can manage your own certificate and store and manipulate credentials. You can find the files behind this functionality in *$SIMPATH/Library/Keychains*.

Network and Proxy Setup

Most of the time, the first step in testing any iOS application is to run it through a proxy so you can examine and potentially modify traffic going from the device to its remote endpoint. Most iOS security testers I know use BurpSuite[1] for this purpose.

---

1. *http://www.portswigger.net*

Bypassing TLS Validation

There's one major catch to running an app under test through a proxy: iOS resolutely refuses to continue TLS/SSL connections when it cannot authenticate the server's certificate, as well it should. This is, of course, the correct behavior, but your proxy-based testing will screech to a halt rather quickly if iOS can't authenticate your proxy's certificate.

For BurpSuite specifically, you can obtain a CA certificate simply by configuring your device or iOS Simulator to use Burp as a proxy and then browsing to *http://burp/cert/* in Mobile Safari. This should work either on a real device or in the iOS Simulator. You can also install CA certificates onto a physical device by either emailing them to yourself or navigating to them on a web server.

For the iOS Simulator, a more general approach that works with almost any web proxy is to add the fingerprint of your proxy software's CA certificate directly into the iOS Simulator trust store. The trust store is a SQLite database, making it slightly more cumbersome to edit than typical certificate bundles. I recommend writing a script to automate this task. If you want to see an example to get you started, Gotham Digital Science has already created a Python script that does the job. You'll find the script here: *https://github.com/GDSSecurity/Add-Trusted-Certificate-to-iOS-Simulator/*.

To use this script, you need to obtain the CA certificate you want to install into the trust store. First configure Firefox[2] to use your local proxy (127.0.0.1, port 8080 for Burp). Then attempt to visit any SSL site; you should get a familiar certificate warning. Navigate to **Add Exception** → **View** → **Details** and click the **PortSwigger CA** entry, as shown in Figure 4-1.

Click **Export** and follow the prompts. Once you've saved the CA certificate, open *Terminal.app* and run the Python script to add the certificate to the store as follows:

```
$ python ./add_ca_to_iossim.py ~/Downloads/PortSwiggerCA.pem
```

Unfortunately, at the time of writing, there isn't a native way to configure the iOS Simulator to go through an HTTP proxy without also routing the rest of your system through the proxy. Therefore, you'll need to configure the proxy in your host system's Preferences, as shown in Figure 4-2.

If you're using the machine for both testing and other work activities, you might consider specifically configuring other applications to go through a separate proxy, using something like FoxyProxy[3] for your browser.

---

2. I generally consider Chrome a more secure daily browser, but the self-contained nature of Firefox does let you tweak proxy settings more conveniently.

3. *http://getfoxyproxy.org*

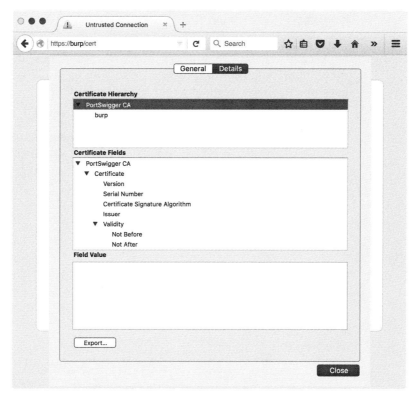

Figure 4-1: Selecting the PortSwigger CA for export

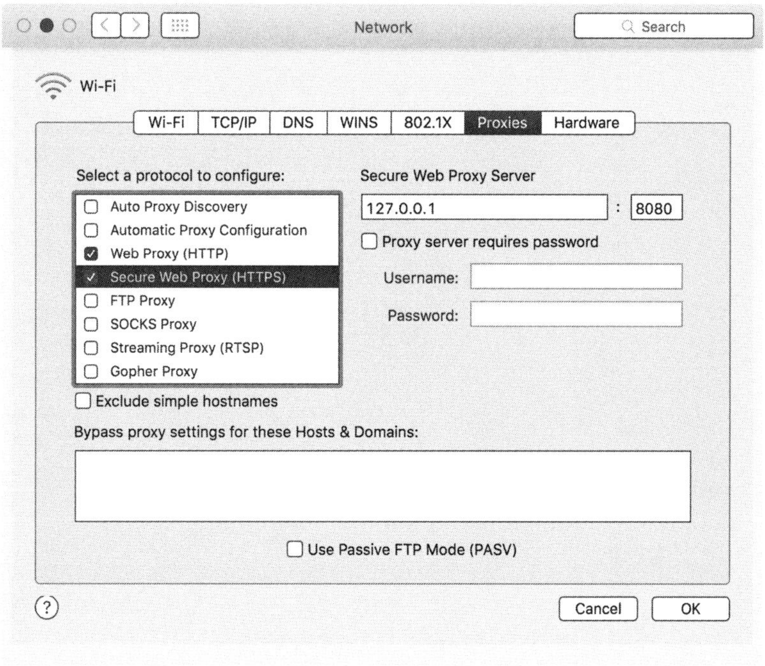

Figure 4-2: Configuring the host system to connect via Burp

Bypassing SSL with stunnel

One method of bypassing SSL endpoint verification is to set up a termination point locally and then direct your application to use that instead. You can often accomplish this without recompiling the application, simply by modifying a plist file containing the endpoint URL.

This setup is particularly useful if you want to observe traffic easily in plaintext (for example, with Wireshark), but the Internet-accessible endpoint is available only over HTTPS. First, download and install stunnel,[4] which will act as a broker between the HTTPS endpoint and your local machine. If installed via Homebrew, stunnel's configuration file will be in */usr/local/etc/stunnel/stunnel.conf-sample*. Move or copy this file to */usr/local/etc/stunnel/stunnel.conf* and edit it to reflect the following:

```
; SSL client mode
client = yes

; service-level configuration
[https]
accept  = 127.0.0.1:80
connect = 10.10.1.50:443
TIMEOUTclose = 0
```

This simply sets up stunnel in client mode, instructing it to accept connections on your loopback interface on port 80, while forwarding them to the remote endpoint over SSL. After editing this file, set up Burp so that it uses your loopback listener as a proxy, making sure to select the **Support invisible proxying** option, as shown in Figure 4-3. Figure 4-4 shows the resulting setup.

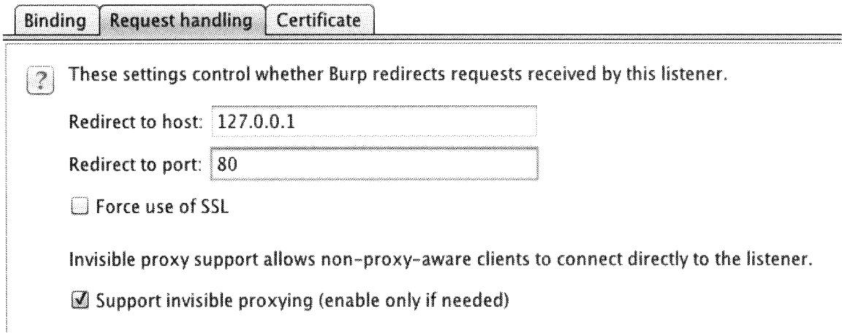

| Binding | Request handling | Certificate |

[?] These settings control whether Burp redirects requests received by this listener.

Redirect to host: 127.0.0.1

Redirect to port: 80

☐ Force use of SSL

Invisible proxy support allows non–proxy–aware clients to connect directly to the listener.

☑ Support invisible proxying (enable only if needed)

Figure 4-3: Setting up invisible proxying through the local stunnel endpoint

---

4. *http://www.stunnel.org/*

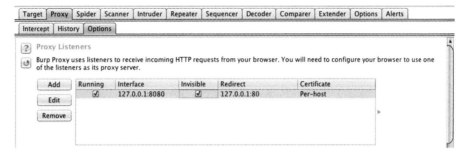

Figure 4-4: Final Burp/stunnel setup

Certificate Management on a Device

To install a certificate on a physical iOS device, simply email the certificate to an account associated with the device or put it on a public web server and navigate to it using Mobile Safari. You can then import it into the device's trust store, as shown in Figure 4-5. You can also configure your device to go through a network proxy (that is, Burp) hosted on another machine. Simply install the CA certificate (as described earlier) of the proxy onto the device and configure your proxy to listen on a network-accessible IP address, as in Figure 4-6.

Figure 4-5: The certificate import prompt

Figure 4-6: Configuring Burp to use a nonlocalhost IP address

Proxy Setup on a Device

Once you've configured your certificate authorities and set up the proxy, go to **Settings** → **Network** → **Wi-Fi** and click the arrow to the right of your currently selected wireless network. You can enter the proxy address and port from this screen (see Figure 4-7).

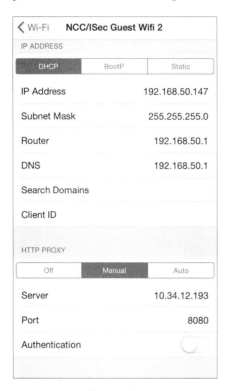

Figure 4-7: Configuring the device to use a test proxy on an internal network

Note that when configuring a device to use a proxy, only connections initiated by NSURLConnection or NSURLSession will obey the proxy settings; other connections such as NSStream and CFStream (which I'll discuss further in Chapter 7) will not. And of course, since this is an HTTP proxy, it works only for HTTP traffic. If you have an application using CFStream, you can edit the codebase with the following code snippet to route stream traffic through the same proxy as the host OS:

```
CFDictionaryRef systemProxySettings = CFNetworkCopySystemProxySettings();

CFReadStreamSetProperty(readStream, kCFStreamPropertyHTTPProxy, systemProxySettings
    );

CFWriteStreamSetProperty(writeStream, kCFStreamPropertyHTTPProxy,
    systemProxySettings);
```

If you're using `NSStream`, you can accomplish the same by casting the `NSInputStream` and `NSOutputStream` to their Core Foundation counterparts, like so:

```
CFDictionaryRef systemProxySettings = CFNetworkCopySystemProxySettings();

CFReadStreamSetProperty((CFReadStreamRef)readStream, kCFStreamPropertyHTTPProxy, (
    CFTypeRef)systemProxySettings);

CFWriteStreamSetProperty((CFWriteStreamRef)writeStream, kCFStreamPropertyHTTPProxy,
    (CFTypeRef)systemProxySettings);
```

If you're doing black-box testing and have an app that refuses to honor system proxy settings for HTTP requests, you can attempt to direct traffic through a proxy by adding a line to */etc/hosts* on the device to point name lookups to your proxy address, as shown in Listing 4-1.

```
10.50.22.11    myproxy api.testtarget.com
```

Listing 4-1: Adding a hosts file entry

You can also configure the device to use a DNS server controlled by you, which doesn't require jailbreaking your test device. One way to do this is to use Tim Newsham's dnsRedir,[5] a Python script that will provide a spoofed answer for DNS queries of a particular domain, while passing on queries for all other domains to another DNS server (by default, 8.8.8.8, but you can change this with the -d flag). The script can be used as follows:

```
$ dnsRedir.py 'A:www.evil.com.=1.2.3.4'
```

This should answer queries for *www.evil.com* with the IP address 1.2.3.4, where that IP address should usually be the IP address of the test machine you're proxying data through.

For non-HTTP traffic, things are a little more involved. You'll need to use a TCP proxy to intercept traffic. The aforementioned Tim Newsham has written a program that can make this simpler—the aptly named tcpprox.[6] If you use the hosts file method in Listing 4-1 to point the device to your proxy machine, you can then have tcpprox dynamically create SSL certificates and proxy the connection to the remote endpoint. To do this, you'll need to create a certificate authority certificate and install it on the device, as shown in Listing 4-2.

---

5. *https://github.com/iSECPartners/dnsRedir/*

6. *https://github.com/iSECPartners/tcpprox/*

```
$ ./prox.py -h
Usage: prox.py [opts] addr port

Options:
  -h, --help     show this help message and exit
  -6             Use IPv6
  -b BINDADDR    Address to bind to
  -L LOCPORT     Local port to listen on
  -s             Use SSL for incoming and outgoing connections
  --ssl-in       Use SSL for incoming connections
  --ssl-out      Use SSL for outgoing connections
  -3             Use SSLv3 protocol
  -T             Use TLSv1 protocol
  -C CERT        Cert for SSL
  -A AUTOCNAME   CName for Auto-generated SSL cert
  -1             Handle a single connection
  -l LOGFILE     Filename to log to

$ ./ca.py -c
$ ./pkcs12.sh ca
  (install CA cert on the device)
$ ./prox.py -s -L 8888 -A ssl.testtarget.com ssl.testtarget.com 8888
```

Listing 4-2: Creating a certificate and using tcpprox to intercept traffic

The *ca.py* script creates the signed certificate, and the *pkcs12.sh* script produces the certificate to install on the device, the same as shown in Figure 4-5. After running these and installing the certificate, your application should connect to the remote endpoint using the proxy, even for SSL connections. Once you've performed some testing, you can read the results with the *proxcat.py* script included with tcpprox, as follows:

```
$ ./proxcat.py -x log.txt
```

Once your application is connected through a proxy, you can start setting up your Xcode environment.

Xcode and Build Setup

Xcode contains a twisty maze of project configuration options—hardly anyone understands what each one does. This section takes a closer look at these options, discusses why you would or wouldn't want them, and shows you how to get Xcode to help you find bugs before they become real problems.

Make Life Difficult

First things first: treat warnings as errors. Most of the warnings generated by clang, Xcode's compiler frontend, are worth paying attention to. Not only do they often help reduce code complexity and ensure correct syntax, they also catch a number of errors that might be hard to spot, such as signedness issues or format string flaws. For example, consider the following:

```
- (void) validate:(NSArray*) someTribbles withValue:(NSInteger) desired {

    if (desired > [someTribbles count]) {
        [self allocateTribblesWithNumberOfTribbles:desired];
    }
}
```

The count method of NSArray returns an unsigned integer, (NSUInteger). If you were expecting the number of desired tribbles from user input, a submitted value might be –1, presumably indicating that the user would prefer to have an anti-tribble. Because desired is an integer being compared to an unsigned integer, the compiler will treat both as unsigned integers. Therefore, this method would unexpectedly allocate an absurd number of tribbles because –1 is an extremely large number when converted to an unsigned integer. I'll discuss this type of integer overflow issue further in Chapter 11.

You can have clang flag this type of of bug by enabling most warnings and treating them as errors, in which case your build would fail with a message indicating "Comparison of integers of different signs: 'int' and 'NSUInteger' (aka 'unsigned int')".

NOTE   *In general, you should enable all warnings in your project build configuration and promote warnings to errors so that you are forced to deal with bugs as early as possible in the development cycle.*

You can enable these options in your project and target build settings. To do so, first, under Warning Policies, set Treat Warnings as Errors to **Yes** (Figure 4-8). Then, under the Warnings sections, turn on all the desired options.

Note that not every build warning that clang supports has an exposed toggle in the Xcode UI. To develop in "hard mode," you can add the -Wextra or -Weverything flag, as in Figure 4-9. Not all warnings will be useful, but it's best to try to understand exactly what an option intends to highlight before disabling it.

-Weverything, used in Figure 4-9, is probably overkill unless you're curious about clang internals; -Wextra is normally sufficient. To save you a bit of time, Table 4-1 discusses two warnings that are almost sure to get in your way (or that are just plain bizarre).

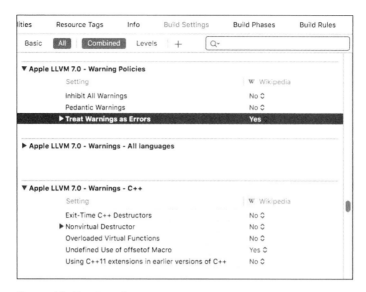

Figure 4-8: Treating all warnings as errors

Figure 4-9: This setting enables all warnings, including options for which there is no exposed UI.

Table 4-1: Obnoxious Warnings to Disable in Xcode

Compiler warning	Justification for disabling
Implicit synthesized properties	Since property synthesis is now automatic, this isn't really an error unless your development guidelines require explicit synthesis.
Unused parameters/functions/variables etc.	These can be supremely irritating when writing code, since your code is obviously not completely implemented yet. Consider enabling these only for nondebug builds.

Enabling Full ASLR

In iOS 4.3, Apple introduced *address space layout randomization (ASLR)*. ASLR ensures that the in-memory structure of the program and its data (libraries, the main executable, stack and heap, and memory-mapped files) are loaded into less predictable locations in the virtual address space. This makes code execution exploits more difficult because many rely on referencing the virtual addresses of specific library calls, as well as referencing data on the stack or heap.

For this to be fully effective, however, the application must be built as a *position-independent executable (PIE)*, which instructs the compiler to build machine code that can function regardless of its location in memory. Without this option, the location of the base executable and the stack will remain the same, even across reboots,[7] making an attacker's job much easier.

To ensure that full ASLR with PIE is enabled, check that Deployment Target in your Target's settings is set to at least iOS version 4.3. In your project's Build Settings, ensure that Generate Position-Dependent Code is set to No and that the bizarrely named Don't Create Position Independent Executable is also set to No. So don't create position-independent executables. Got it?

For black-box testing or to ensure that your app is built with ASLR correctly, you can use otool on the binary, as follows:

```
$ unzip MyApp.ipa
$ cd Payload/MyApp.app
$ otool -vh MyApp

MyApp (architecture armv7):
Mach header
      magic cputype cpusubtype caps   filetype ncmds sizeofcmds        flags
   MH_MAGIC    ARM         V7 0x00     EXECUTE    21       2672 NOUNDEFS DYLDLINK
                                                                TWOLEVEL PIE
```

7. *http://www.trailofbits.com/resources/ios4_security_evaluation_paper.pdf*

```
MyApp (architecture armv7s):
Mach header
      magic cputype cpusubtype caps   filetype ncmds sizeofcmds         flags
   MH_MAGIC    ARM       V7S 0x00    EXECUTE    21       2672 NOUNDEFS DYLDLINK
                                                               TWOLEVEL PIE
```

At the end of each MH_MAGIC line, if you have your settings correct, you should see the PIE flag, highlighted in bold. (Note that this must be done on a binary compiled for an iOS device and will not work when used on iOS Simulator binaries.)

Clang and Static Analysis

In computer security, *static analysis* generally refers to using tools to analyze a codebase and identify security flaws. This could involve identifying dangerous APIs, or it might include analyzing data flow through the program to identify the potentially unsafe handling of program inputs. As part of the build tool chain, clang is a good spot to embed static analysis language.

Beginning with Xcode 3.2, clang's static analyzer[8] has been integrated with Xcode, providing users with a UI to trace logic, coding flaws, and general API misuse. While clang's static analyzer is handy, several of its important features are disabled by default in Xcode. Notably, the checks for classic dangerous C library functions, such as strcpy and strcat, are oddly absent. Enable these in your Project or Target settings, as in Figure 4-10.

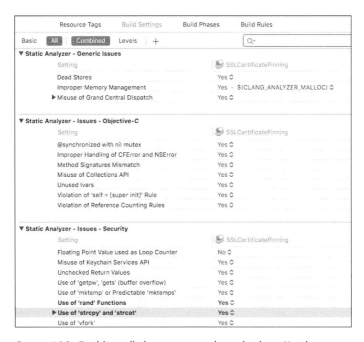

Figure 4-10: Enabling all clang static analysis checks in Xcode

8. *http://clang-analyzer.llvm.org/*

Address Sanitizer and Dynamic Analysis

Recent versions of Xcode include a version of clang/llvm that features the Address Sanitizer (ASan). ASan is a dynamic analysis tool similar to Valgrind, but ASan runs faster and has improved coverage.[9] ASan tests for stack and heap overflows and use-after-free bugs, among other things, to help you track down crucial security flaws. It does have a performance impact (program execution is estimated to be roughly two times slower), so don't enable it on your release builds, but it should be perfectly usable during testing, quality assurance, or fuzzing runs.

To enable ASan, add `-fsanitize=address` to your compiler flags for debug builds (see Figure 4-11). On any unsafe crashes, ASan should write extra debug information to the console to help you determine the nature and severity of the issues. In conjunction with fuzzing,[10] ASan can be a great help in pinning down serious issues that may be security-sensitive and in giving an idea of their exploitability.

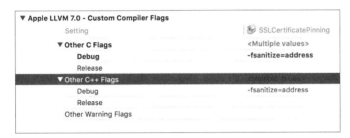

Figure 4-11: Setting the ASan compiler flags

Monitoring Programs with Instruments

Regardless of whether you're analyzing someone else's application or trying to improve your own, the DTrace-powered Instruments tool is extremely helpful for observing an app's activity on a fine-grained level. This tool is useful for monitoring network socket usage, finding memory allocation issues, and watching filesystem interactions. Instruments can be an excellent tool for discovering what objects an application stores on local storage in order to find places where sensitive information might leak; I use it in that way frequently.

Activating Instruments

To use Instruments on an application from within Xcode, hold down the **Run** button and select the **Build for Profiling** option (see Figure 4-12). After building, you will be presented with a list of preconfigured templates tailored for monitoring certain resources, such as disk reads and writes, memory allocations, CPU usage, and so on.

---

9. *http://clang.llvm.org/docs/AddressSanitizer.html*

10. *http://blog.chromium.org/2012/04/fuzzing-for-security.html*

Figure 4-12: Selecting the Build for Profiling option

The File Activity template (shown in Figure 4-13) will help you monitor your application's disk I/O operations. After selecting the template, the iOS Simulator should automatically launch your application and begin recording its activity.

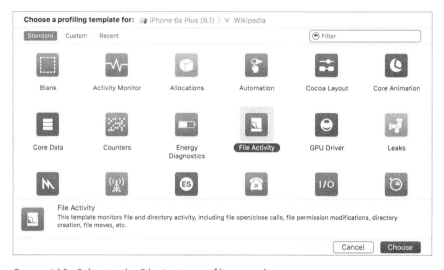

Figure 4-13: Selecting the File Activity profiling template

There are a few preset views in Instruments for monitoring file activity. A good place to start is Directory I/O, which will capture all file creation or deletion events. Test your application the way you normally would and watch the output here. Each event is listed with its Objective-C caller, the C function call underlying it, the file's full path, and its new path if the event is a rename operation.

You'll likely notice several types of cache files being written here (see Figure 4-14), as well as cookies or documents your application has been asked to open. If you suspend your application, you should see the application screenshot written to disk, which I'll discuss in Chapter 10.

For a more detailed view, you can select the Reads/Writes view, as shown in Figure 4-15. This will show any read or write operations on files or sockets, along with statistics on the amount of data read or written.

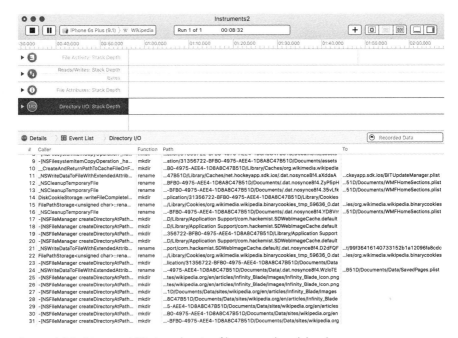

Figure 4-14: Directory I/O view showing files created or deleted

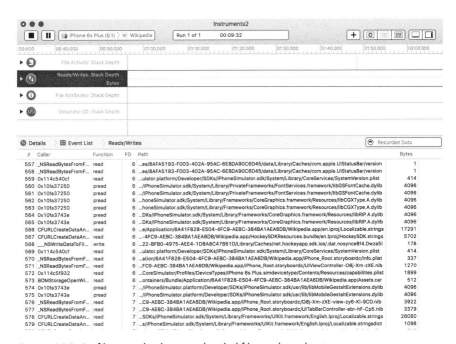

Figure 4-15: Profiling results showing detailed file reads and writes

Watching Filesystem Activity with Watchdog

Instruments should catch most iOS Simulator activity, but some file writes or network calls may actually be performed by other system services, thereby escaping the tool's notice. It's a good idea to manually inspect the iOS Simulator's directory tree to get a better feel for the structure of iOS and its applications and to catch application activity that you might otherwise miss.

One easy way to automate this is to use the Python watchdog module.[11] Watchdog will use either the kqueue or FSEvents API to monitor directory trees for file activity and can either log events or take specific actions when these events occur. To install watchdog, use the following:

```
$ pip install watchdog
```

You can write your own scripts to use watchdog's functionality, but you'll find a nice command line tool already included with watchdog called watchmedo. If you open a Terminal window and navigate to the Simulator directory, you should be able to use watchmedo to monitor all file changes under the iOS Simulator's directory tree, as follows:

```
$ cd ~/Library/Application\ Support/iPhone\ Simulator/6.1
$ watchmedo log --recursive .
on_modified(self=<watchdog.tricks.LoggerTrick object at 0x103c9b190>, event=<
    DirModifiedEvent: src_path=/Users/dthiel/Library/Application Support/iPhone
    Simulator/6.1/Library/Preferences>)
on_created(self=<watchdog.tricks.LoggerTrick object at 0x103c9b190>, event=<
    FileCreatedEvent: src_path=/Users/dthiel/Library/Application Support/iPhone
    Simulator/6.1/Applications/9460475C-B94A-43E8-89C0-285DD036DA7A/Library/Caches
    /Snapshots/com.yourcompany.UICatalog/UIApplicationAutomaticSnapshotDefault-
    Portrait.png>)
on_modified(self=<watchdog.tricks.LoggerTrick object at 0x103c9b190>, event=<
    DirModifiedEvent: src_path=/Users/dthiel/Library/Application Support/iPhone
    Simulator/6.1/Applications/9460475C-B94A-43E8-89C0-285DD036DA7A/Library/Caches
    /Snapshots>)
on_created(self=<watchdog.tricks.LoggerTrick object at 0x103c9b190>, event=<
    DirCreatedEvent: src_path=/Users/dthiel/Library/Application Support/iPhone
    Simulator/6.1/Applications/9460475C-B94A-43E8-89C0-285DD036DA7A/Library/Caches
    /Snapshots/com.yourcompany.UICatalog>)
on_modified(self=<watchdog.tricks.LoggerTrick object at 0x103c9b190>, event=<
    DirModifiedEvent: src_path=/Users/dthiel/Library/Application Support/iPhone
    Simulator/6.1/Library/SpringBoard>)
```

---

11. *https://pypi.python.org/pypi/watchdog/*

Entries that start with on `on_modified` indicate a file was changed, and entries that start with `on_created` indicate a new file. There are several other change indicators you might see from watchmedo, and you can read about them in the Watchdog documentation.

Closing Thoughts

You should now have your build and test environment configured for running, modifying, and examining iOS apps. In Chapter 5, we'll take a closer look at how to debug and inspect applications dynamically, as well as how to change their behavior at runtime.

5

DEBUGGING WITH LLDB
AND FRIENDS

Debugging iOS applications is considered one of
Xcode's strong components. In addition to the useful
analysis features of DTrace, Xcode has a command
line debugger with a relatively approachable graphical interface. As part of Apple's migration away from
GNU utilities, the default debugger is now lldb,[1] which
provides first-class support for Objective-C. Multithreaded debugging is well-
supported, and you can even inspect objects from the debugger. The only
downside is that you'll have to translate your hard-won knowledge of gdb to
a new environment.

Debugging is a vast topic, and there are multiple books on the subject.[2]
This chapter covers the basics for people new to Xcode, along with tips
relevant to security testing and secure development. I assume you have some
familiarity with gdb and debuggers in general.

---

1. *http://lldb.llvm.org/*

2. For a detailed resource on debugging in Xcode, I recommend *iOS 7 Programming: Pushing the
Limits*; see *http://iosptl.com/*.

Useful Features in lldb

Xcode's built-in debugger interface is fairly powerful. It has a command line, but you can also use the GUI to view and interact with the current thread state, annotated assembly, and object details. The GUI includes a central breakpoint browser as well, where you can view, enable, and disable breakpoints.

NOTE   *If you're extremely comfortable using gdb, the LLVM project has a mapping of frequently used gdb commands to their lldb equivalents; see* http://lldb.llvm.org/lldb-gdb.html.

Working with Breakpoints

You can set breakpoints graphically from Xcode's lldb interface (see Figure 5-1), or you can do so from the command line. In addition to breaking when the program accesses a particular memory address or C function, you can also break on specific Objective-C methods.

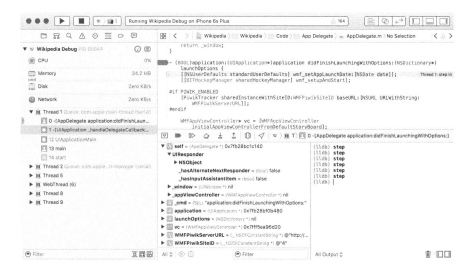

Figure 5-1: Xcode's lldb interface

Here are some of the ways you can set breakpoints:

---

❶ (lldb) breakpoint set --name myfunction --name myotherfunction
❷ (lldb) breakpoint set --name "-[myClass methodCall:]"
❸ (lldb) breakpoint set --selector myObjCSelector:
❹ (lldb) breakpoint set --method myCPlusPlusMethod

The command at ❶ sets one breakpoint on multiple functions, a feature you can use to enable and disable groups of functions simultaneously. As shown at ❷, you can also break on specific Objective-C instance and class methods—these can be also be grouped in a manner similar to the C function calls at ❶. If you want to break on all calls to a particular selector/method, use the --selector option ❸, which will break on any calls to a selector of this name, regardless of what class they're implemented in. Finally, to break on specific C++ methods, simply specify --method instead of --name when defining the breakpoint, as at ❹.

In practice, setting a breakpoint in lldb looks like this:

```
(lldb) breakpoint set --name main
Breakpoint 2: where = StatePreservator`main + 34 at main.m:15, address = 0x00002822

(lldb) breakpoint set -S encodeRestorableStateWithCoder:
Breakpoint 2: where = StatePreservator`-[StatePreservatorSecondViewController
    encodeRestorableStateWithCoder:] + 44 at StatePreservatorSecondViewController.
    m:25, address = 0x00002d5c
```

After you set a breakpoint, lldb shows the code you're breaking on. If you like, you can make this even simpler: like gdb, lldb recognizes keywords using the shortest matching text. So *breakpoint* can be shortened to *break*, or even *b*.

In the GUI, you can break on a particular line of code by clicking the number in the gutter to the left of the line (see Figure 5-2). Clicking again will disable the breakpoint. Alternatively, you can break on lines from the lldb CLI using the --file *filename.m* --line *66* syntax.

```
316
317    NSIndexPath *selectedIndexPath = [self.tableView indexPathForSelectedRow];
318    if (selectedIndexPath.section == 0)
319        numToPreview = NUM_DOCS;
320    else
321        numToPreview = self.documentURLs.count;
322
323    return numToPreview;
```

Figure 5-2: Setting breakpoints on specific lines with the mouse. Deactivated breakpoints are shaded a lighter gray.

When you want to create multiple breakpoints, it can be handy to use the -r flag at the command line to break on functions matching a particular regular expression, like so:

```
(lldb) break set -r tableView
Breakpoint 1: 4 locations.
(lldb) break list
Current breakpoints:
1: source regex = "tableView", locations = 4, resolved = 4
  1.1: where = DocInteraction`-[DITableViewController tableView:
     cellForRowAtIndexPath:] + 695 at DITableViewController.m:225, address = 0
     x000032c7, resolved, hit count = 0
  1.2: where = DocInteraction`-[DITableViewController tableView:
     cellForRowAtIndexPath:] + 1202 at DITableViewController.m:245, address = 0
     x000034c2, resolved, hit count = 0
```

```
1.3: where = DocInteraction`-[DITableViewController tableView:
    cellForRowAtIndexPath:] + 1270 at DITableViewController.m:246, address = 0
    x00003506, resolved, hit count = 0
1.4: where = DocInteraction`-[DITableViewController tableView:
    cellForRowAtIndexPath:] + 1322 at DITableViewController.m:247, address = 0
    x0000353a, resolved, hit count = 0
```

This will set a single breakpoint with a number of *locations*. Each location can be enabled and disabled, as shown here:

```
(lldb) break dis 1.4
1 breakpoints disabled.
(lldb) break list
Current breakpoints:
1: source regex = ".*imageView.*", locations = 4, resolved = 3
    --snip--
  1.4: where = DocInteraction`-[DITableViewController tableView:
      cellForRowAtIndexPath:] + 1322 at DITableViewController.m:247, address = 0
      x0000353a, unresolved, hit count = 0  Options: disabled
(lldb) break en 1.4
1 breakpoints disabled.
```

Notice that enabling and disabling locations works just like a regular breakpoint; just use break disable and break enable and reference the right numeric identifier.

Navigating Frames and Variables

Once you've arrived at a breakpoint, you can use lldb to examine the state of your program. You can do this via either the command line, as in the other lldb examples I've shown, or the visual lldb browser, as in Figure 5-3.

Figure 5-3: Examining frame variables from the command line and the GUI

In addition to viewing and manipulating the variables of the current frame, you can navigate the program threads and frames of the call stack using the Debug Navigator, as shown in Figure 5-4.

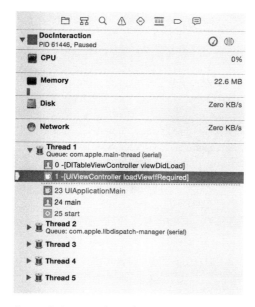

Figure 5-4: Using the Debug Navigator to switch frames and threads

Similar to using gdb, you can inspect the call stack of the current thread with the bt (short for *backtrace*) command (see Listing 5-1). Normally, you could also navigate frames using the typical up, down, and frame select commands. In some versions of Xcode however, a bug causes the frame to immediately revert to the frame selected in the Debug Navigator. In that case, you must switch frames manually within the Debug Navigator to inspect them individually.

```
(lldb) bt
* thread #1: tid = 0x11804c, 0x00002c07 StatePreservator`-[
    StatePreservatorSecondViewController encodeRestorableStateWithCoder:](self=0
    x07733c30, _cmd=0x005af437, coder=0x0756faf0) + 55 at
    StatePreservatorSecondViewController.m:25, queue = 'com.apple.main-thread,
    stop reason = breakpoint 1.1
  frame #0: 0x00002c07 StatePreservator`-[StatePreservatorSecondViewController
    encodeRestorableStateWithCoder:](self=0x07733c30, _cmd=0x005af437, coder=0
    x0756faf0) + 55 at StatePreservatorSecondViewController.m:25
  frame #1: 0x000277e7 UIKit`-[UIApplication(StateRestoration)
    _saveApplicationPreservationState:] + 1955
  frame #2: 0x00027017 UIKit`-[UIApplication(StateRestoration)
    _saveApplicationPreservationStateIfSupported] + 434
  frame #3: 0x0001b07b UIKit`-[UIApplication _handleApplicationSuspend:eventInfo
    :] + 947
```

```
frame #4: 0x00023e74 UIKit`-[UIApplication handleEvent:withNewEvent:] + 1469
frame #5: 0x00024beb UIKit`-[UIApplication sendEvent:] + 85
frame #6: 0x00016698 UIKit`_UIApplicationHandleEvent + 9874
frame #7: 0x01beddf9 GraphicsServices`_PurpleEventCallback + 339
frame #8: 0x01bedad0 GraphicsServices`PurpleEventCallback + 46
frame #9: 0x01c07bf5 CoreFoundation`
  __CFRUNLOOP_IS_CALLING_OUT_TO_A_SOURCE1_PERFORM_FUNCTION__ + 53
frame #10: 0x01c07962 CoreFoundation`__CFRunLoopDoSource1 + 146
frame #11: 0x01c38bb6 CoreFoundation`__CFRunLoopRun + 2118
frame #12: 0x01c37f44 CoreFoundation`CFRunLoopRunSpecific + 276
frame #13: 0x01c37e1b CoreFoundation`CFRunLoopRunInMode + 123
frame #14: 0x01bec7e3 GraphicsServices`GSEventRunModal + 88
frame #15: 0x01bec668 GraphicsServices`GSEventRun + 104
frame #16: 0x00013ffc UIKit`UIApplicationMain + 1211
frame #17: 0x0000267d StatePreservator`main(argc=1, argv=0xbffff13c) + 141 at
  main.m:16
```

Listing 5-1: Getting the current call stack with the backtrace *command*

To examine the variables of the current frame, you can use the `frame variable` command, as shown in Listing **??**.

```
(lldb) frame variable
(StatePreservatorSecondViewController *const) self = 0x0752d2e0
(SEL) _cmd = "encodeRestorableStateWithCoder:"
(NSCoder *) coder = 0x0d0234e0
```

Listing 5-2: Using the frame variable *command*

This will give you variable names and arguments of the local stack frame, along with their types and memory addresses. You can also use the context menu in the graphical debugger to print or edit variable contents; see Figure 5-5.

If you use `frame select` on its own, you can also see the program's location in the call stack, along with the relevant surrounding lines of code, as in this example:

```
(lldb) frame select
frame #0: 0x00002d5c StatePreservator`-[StatePreservatorSecondViewController
    encodeRestorableStateWithCoder:](self=0x0752d2e0, _cmd=0x005af437, coder=0
    x0d0234e0) + 44 at StatePreservatorSecondViewController.m:25
   22
   23  -(void)encodeRestorableStateWithCoder:(NSCoder *)coder
   24  {
-> 25      [coder encodeObject:[_myTextView text] forKey:@"UnsavedText"];
   26      [super encodeRestorableStateWithCoder:coder];
   27  }
   28
```

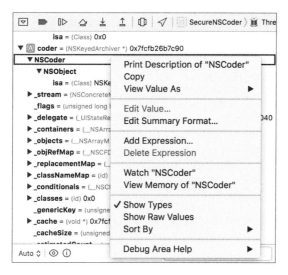

Figure 5-5: The variable context menu, showing options for printing variable contents, setting watchpoints, and viewing memory contents

The frame select command also takes a numeric argument for the stack frame you want to inspect, if you'd like to look further up the call stack (see Listing 5-3).

```
(lldb) frame select 4
frame #4: 0x00023e74 UIKit`-[UIApplication handleEvent:withNewEvent:] + 1469
UIKit`-[UIApplication handleEvent:withNewEvent:] + 1469:
-> 0x23e74:  xorb    %cl, %cl
   0x23e76:  jmp     0x24808
                     ; -[UIApplication handleEvent:withNewEvent:] + 3921
   0x23e7b:  movl    16(%ebp), %ebx
   0x23e7e:  movl    %ebx, (%esp)
```

Listing 5-3: Assembly shown while examining a stack frame

Note that for code outside of your current project, such as other parts of the Cocoa API, the source will usually not be available; lldb will instead show you the relevant assembly instructions.[3]

You can also inspect the values of objects using lldb's po (short for *print object*) command. For example, consider the following:

```
(lldb) po [self window]
$2 = 0x071848d0 <UIWindow: 0x71848d0; frame = (0 0; 320 480); hidden = YES; layer =
      <UIWindowLayer: 0x71849a0>>
```

---

3. If you'd like further insight into assembly on iOS and ARM, check out Ray Wenderlich's tutorial at *http://www.raywenderlich.com/37181/ios-assembly-tutorial/*.

Using po on your main window fetches the addresses and attributes of that window.

Visually Inspecting Objects

If you're using Xcode 5 or later, you can also hover the mouse over objects to inspect the contents, as shown in Figure 5-6. If you drill down into individual subobjects, you can either view their memory directly (Figure 5-7) by clicking the **i** button or use the Quick Look "eye" button to see the contents of the object represented as a fully rendered image, text, or any other data type that OS X's Quick Look API understands (see Figure 5-8). This is, in my opinion, pretty badass.

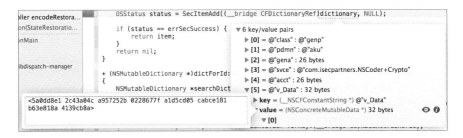

Figure 5-6: Inspecting an object while at a breakpoint

Figure 5-7: Inspecting an object's contents in memory

Figure 5-8: Examining the current state of a variable with the Quick Look button. In this case, you're looking at the _statusBar of the UIApplication delegate window, which Xcode will display as an actual image.

Manipulating Variables and Properties

You can do more than just view the contents of variables and objects from lldb. For example, let's try breaking on the same line used to test the frame variable command back in Listing 5-2.

```
[coder encodeObject:[_myTextView text] forKey:@"UnsavedText"];
```

When the debugger reaches this line, imagine you want to examine the contents of UITextView's *text* attribute and change its value before the program continues. You can do this with the expr command, using traditional Objective-C syntax, as follows:

```
(lldb) po [_myTextView text]
$0 = 0x08154cb0 Foo
(lldb) expr (void)[_myTextView setText:@"Bar"]
(lldb) po [_myTextView text]
$1 = 0x0806b2e0 Bar
(lldb) cont
```

When execution resumes, the value of that text box in the UI should have changed. Because lldb doesn't know the return type of a method called in this way, you have to specify the type using (void) with the expr command. Similarly, if you were calling something that returned an int, you'd need to explicitly cast to that type instead. For simple assignment operations, like myInteger = 666 or similar, as opposed to method calls, simply enter expr and the assignment as one command.

NOTE   *When using lldb from the command line in Xcode, the GUI will autocomplete object method names, giving you a brief description and their return type. See Figure 5-9 for an example.*

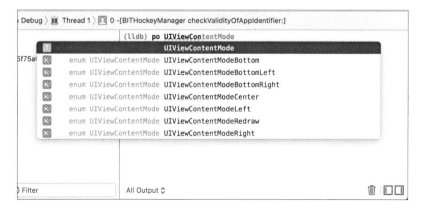

Figure 5-9: Nifty lldb method name completion in Xcode

Keep in mind that you're not limited to manipulating objects that are declared in your code. You can also manipulate framework classes.

```
(lldb) expr (void)[[UIPasteboard generalPasteboard] setString:@"my string"]
(lldb) po [[UIPasteboard generalPasteboard] string]
$5 = 0x071c6e50 my string
```

For this kind of interactive manipulation and interrogation, I often find it useful to set a breakpoint on didReceiveMemoryWarning in the application delegate because this method will be present in every application. When I want to inspect the program's state while running it in the iOS Simulator, I select Hardware → Simulate Memory Warning. Once I've done my twiddling, I simply continue the application with cont. You can also do this from the Xcode UI with the Pause Execution button.

Breakpoint Actions

Breakpoint actions are not well-documented but are quite useful. They allow you to create breakpoints that trigger only under certain conditions, and they can perform complex actions when these breakpoints are hit. You can set them up to automatically resume execution after performing these actions or even have them trigger only after a line is hit a certain number of times. Logging and using speech synthesis to present program information are the simplest actions you can set for a breakpoint, but you can also interrogate objects, read and manipulate variables, and so forth. Basically, breakpoint actions can do anything you can do from the lldb command line, plus a few other niceties.

Let's walk through creating a breakpoint action one step at a time.

1. Create a breakpoint by clicking in the breakpoint gutter.

2. CTRL-click the breakpoint and select **Edit Breakpoint**.

3. Click **Add Action**.

4. Check the **Automatically continue after evaluating** box.

5. For the simplest type of breakpoint action, simply select the **Log message** action. Here, you can print simple messages, along with the breakpoint name and hit count (see Figure 5-10). You can ignore the expression option because it's not terribly straightforward to use.

6. After adding a simple log message, you can click the + button to add another action. This time, select **Debugger Command**.

7. Here, you can enter basic lldb expressions—most commonly, using the po command to print the description of an object. See Figure 5-11 for an example.

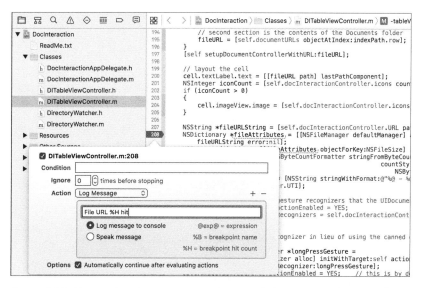

Figure 5-10: Using a breakpoint action to do a simple log entry. In this example, you'll log a message, along with the number of times the breakpoint has been hit, using the %H placeholder.

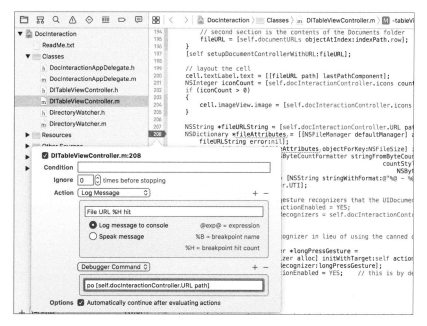

Figure 5-11: In addition to simply logging, you can execute an arbitrary lldb command. In this case, you'll use the po command to print the description of the object returned by the path method.

8. Optionally, add a breakpoint condition to specify when the actions you've defined are executed (Figure 5-12).

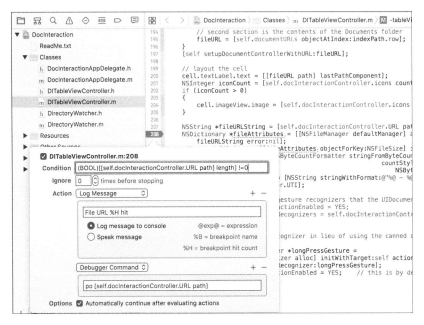

Figure 5-12: Two actions and a breakpoint condition. For the condition, you'll ensure that the length of the path is not zero before executing the breakpoint action, specifying the return value (BOOL).

Try following these steps until you feel comfortable using breakpoint actions, and then move on to the next section for some specific ways to apply lldb in a security context.

Using lldb for Security Analysis

These are all useful tricks, but how do you put them together to find new security issues or test security assertions? Let's take a look at a couple scenarios where using the debugger can help you nail down more concrete issues.

Fault Injection

Say you have an application that uses a custom binary network protocol to marshal data between the client and a remote server. This can make it difficult to intercept and modify data with an off-the-shelf proxy, but you'd like to determine whether malformed data in certain parameters could cause a program to crash. You can also manipulate data to make future testing easier.

Since you can change data, you might want to replace, for example, a randomly generated key with one of your choosing. You can do that from within the debugger, as shown in Listing 5-4. This results in data being encrypted with a known key of your choosing, rather than a potentially unprintable blob. The following example modifies the app's crypto key before it gets saved to the Keychain so that further communication uses a different key:

```
❶ (lldb) frame var
   (Class) self = SimpleKeychainWrapper
   (SEL) _cmd = "addToKeychain:forService:"
   (NSString *) identifier = 0x00005be4 @"com.isecpartners.CryptoKey"
   (NSString *) service = 0x00005bf4 @"com.isecpartners.NSCoder+Crypto"
   (NSMutableDictionary *) dictionary = 0x08b292f0 6 key/value pairs
   (NSMutableData *) item = 0x08b2cee0
   (OSStatus) status = 1
❷ (lldb) po item
   <9aab766a 260bb165 57675f04 fdb982d3 d73365df 5fd4b05f 3c078f7b b6484b7d>
❸ (lldb) po dictionary
   {
       acct = <636f6d2e 69736563 70617274 6e657273 2e437279 70746f4b 6579>;
       class = genp;
       gena = <636f6d2e 69736563 70617274 6e657273 2e437279 70746f4b 6579>;
       pdmn = aku;
       svce = "com.isecpartners.NSCoder+Crypto";
       "v_Data" = <9aab766a 260bb165 57675f04 fdb982d3 d73365df 5fd4b05f 3c078f7b
         b6484b7d>;
   }
❹ (lldb) expr (void)[dictionary setObject:@"mykey" forKey:(__bridge id)kSecValueData
       ];
❺ (lldb) po dictionary
   {
       acct = <636f6d2e 69736563 70617274 6e657273 2e437279 70746f4b 6579>;
       class = genp;
       gena = <636f6d2e 69736563 70617274 6e657273 2e437279 70746f4b 6579>;
       pdmn = aku;
       svce = "com.isecpartners.NSCoder+Crypto";
       "v_Data" = mykey;
   }
```

Listing 5-4: Inspecting and changing object values in memory

At ❶, the code prints the variables of the current frame, noting the arguments sent to the addToKeychain:forService: selector. The key this example is interested in is stored in the item argument and added to a dictionary. Inspecting these (❷ and ❸) reveals the value of the key. The code then alters the Keychain dictionary using the expr command ❹. At ❺, the program verifies that the new NSString is now the current value of the key.

Tracing Data

If you have an application that encrypts data with a master password, it may be useful to examine that data before it gets encrypted. It may not always be immediately obvious that data will hit the encryption routine by default. Consider Listing 5-5:

```
❶ (lldb) frame variable
   (CCCryptHelper *const) self = 0x07534b40
❷ (SEL) _cmd = "encrypt:"
❸ (NSString *) data = 0x0000c0ec @"PasswordManager"
   (NSData *) encData = 0x07534b40 0 byte

   (lldb) frame select
   frame #0: 0x00006790 PasswordManager `-[CCCryptHelper encrypt:](self=0x07534b40,
   _cmd=0x00009b1e, data=0x0000c0ec) + 48 at CCCryptHelper.m:82
   80 - (NSData *)encrypt:(NSString *)data {
   -> 81 NSData *encData = [self AES128EncryptData:[data dataUsingEncoding:
         NSUTF8StringEncoding]
   82                                          withKey:masterPassword];
```

Listing 5-5: Examining frame variables with lldb

If you break on the encrypt: selector ❷, you can examine the local variables using the frame variable command ❶. Notice that the output shows both data and encData. The former ❸ is the interesting bit in this example, because that's the data that will be encrypted and returned by the routine. This tracing technique can also be used to examine and manipulate data to be sent over the wire, before it hits the encryption routines.

Examining Core Frameworks

lldb is also useful for digging in to the weird quirks of Apple's APIs—I recommend you use it when you're confused by an API's behavior. For instance, when looking into NSURLCache, I noticed the behavior in Listing 5-6:

```
(lldb) expr (int)[[NSURLCache sharedURLCache] currentMemoryUsage]
(int) $0 = 158445

(lldb) expr (int)[[NSURLCache sharedURLCache] currentDiskUsage]
(int) $1 = 98304

❶ (lldb) expr (void)[[NSURLCache sharedURLCache] removeAllCachedResponses]

(lldb) expr (int)[[NSURLCache sharedURLCache] currentMemoryUsage]
(int) $3 = 0
```

```
(lldb) expr (int)[[NSURLCache sharedURLCache] currentDiskUsage]
```
❷ `(int) $4 = 98304`

Listing 5-6: Some curious behavior of the NSURLCache *API*

Here, even though I called the removeAllCachedResponses method ❶, the
current disk usage is still 98304 bytes ❷. Alas, it appears that clearing the
cache is useless. Fear not—you'll see some solutions to this problem in
Chapter 9. In the meantime, you may want to play around with some of
the internals yourself. This can help you figure out some of the workings of
the iOS platform and give you deeper insight into how your application is
behaving.

Closing Thoughts

All of these debugging and inspection techniques can be useful when trying to debug your own application or understand a new codebase quickly.
However, you may not always have access to the source code of the product
you're working with. In these cases, you'll want to know some basic black-
box testing techniques, which I will cover in Chapter 6.

6

BLACK-BOX TESTING

While white-box testing is almost always the best way to security test an application, sometimes you simply have to do your testing without source code or insight into a program's design. In these cases, you'll need to dig a little deeper into the guts of iOS, especially into the realm of Objective-C and the Mach-O binary format.

Black-box testing on iOS is a rapidly moving target—it relies on the continuous development of jailbreaks, as well as robust third-party tools and debugging implements. I've tried to make the techniques and tools described in this chapter as future-proof as possible to give you a solid foundation to build on.

To effectively black-box test an iOS application, you'll first need to get a jailbroken device so that you can sideload applications and install your testing tool chain. The details of jailbreaking change too rapidly for me to document here, but you can usually find current information from the iPhone Dev Team[1] or iClarified.[2]

Once you've jailbroken your device, launch Cydia, choose **Developer** mode, and then update your package list (under Changes).

---

1. *http://blog.iphone-dev.org/*

2. *http://iclarified.com/*

Now you can load your device with some testing tools, primarily from the Cydia app store. These are the must-haves:

odcctools This includes otool, lipo, and other development goodies.

OpenSSH You'll need this to actually access the device. Be sure to change the passwords of your root and mobile accounts *immediately* using the passwd(1) command.

MobileTerminal This will allow you to navigate the command line on the device itself, when necessary.

cURL You'll want this for downloading remote files over HTTP or FTP.

Erica Utilities This includes a smattering of useful utilities from Erica Sadun. See a detailed list at *http://ericasadun.com/ftp/EricaUtilities/*.

vbindiff This is a binary diff program to help verify changes to binaries.

netcat This is your general, all-purpose network listener.

rsync You can install this for syncing whole directory trees to and from the device.

tcpdump You can install this for capturing network traffic dumps for analysis.

IPA Installer Console This will allow you to directly install *.ipa* files copied to the device.

Cydia Substrate This tool is used for hooking and modifying the behavior of applications.

Now, let's look at how you can get these testing tools onto your device.

Installing Third-Party Apps

Depending on how you've come to possess your application files, there are a couple of ways to sideload them onto your device.

Using a .app Directory

If you've acquired a *.app* directory, you can do the following:

First, archive your *.app* bundle with tar, and use scp to copy the archive over to your test device, as follows:

```
$ tar -cvzf archive.tar.gz mybundle.app}
$ scp archive.tar.gz root@dev.ice.i.p:
```

Then ssh to your device and untar the bundle into the */Applications* directory:

```
$ cd /Applications
$ tar -xvzf ~/archive.tar.gz
```

This should put the application right next to the official Apple-supplied applications. To get it to show up on the home screen, you'll need to either restart the SpringBoard or reboot the device. To restart SpringBoard, you can use the `killall` command, like this:

```
$ killall -HUP SpringBoard
```

If you find yourself needing to "respring" a lot, you can use a tool like CCRespring from Cydia, as shown in Figure 6-1.

Respring

Figure 6-1: A simple respring button added to the Control Center by CCRespring

Tools like CCRespring add a button that you can press to restart the SpringBoard so you don't have to go to the command line every time.

NOTE   *Some have reported that simply respringing the device does not cause the application to appear on the SpringBoard. In this case, you can either reboot or run the `uicache` command as the mobile user.*

Using a .ipa Package File

If you've been given (or have otherwise obtained) a *.ipa* package file, you can copy it to your device with scp and install it using the installipa command, as follows:

```
$ installipa ./Wikipedia-iOS.ipa
Analyzing Wikipedia-iOS.ipa...
Installing Wikipedia (v3.3)...
Installed Wikipedia (v3.3) successfully.

$ ls Applications/CC189021-7AD0-498F-ACB6-356C9E521962
Documents  Library  Wikipedia-iOS.app  tmp
```

Decrypting Binaries

Before you can inspect the contents of binaries, you'll need to decrypt them. There are a couple of ways to do so. The simplest way is to use a prepackaged tool, such as Stefan Esser's dumpdecrypted.[3] This is a shared library that is dynamically loaded when executing your application. You can use it as follows:

```
$ git clone https://github.com/stefanesser/dumpdecrypted
$ cd dumpdecrypted
$ make
$ scp dumpdecrypted.dylib root@your.dev.ice:
$ ssh root@your.dev.ice
$ DYLD_INSERT_LIBRARIES=dumpdecrypted.dylib /var/mobile/Applications/(APP_ID)/
    YourApp.app/YourApp
```

This will output a decrypted version of the binary within the *tmp* directory of the application's *.app* bundle.

Because there have been many automated tools for dumping decrypted binaries, most of which have become unusable, it's best to have a backup method. For a more robust and (ideally) future-proof way to decrypt binaries and to help you understand some of the inner workings of application encryption and decryption, you can use command line tools and lldb.[4]

To create a decrypted binary, you'll follow these basic steps:

1. Analyze the binary to determine the location of its encrypted portion.

2. Run the application under lldb.

---

3. *https://github.com/stefanesser/dumpdecrypted*

4. Traditionally, this has been done with the GNU Debugger, gdb. However, gdb hasn't been included with Xcode since version 4, and most versions in Cydia are broken. This method of using lldb should work for the foreseeable future . . . I think.

3. Dump the unencrypted segment to disk.

4. Copy the original binary for use as a donor file.

5. Remove the donor binary's cryptid flag.

6. Transplant the unencrypted segment into the donor binary.

Let's discuss this decryption process in more detail.

Launching the debugserver on the Device

Before you can get a memory dump, you need to get Apple's debugserver onto the device. The debugserver is in *DeveloperDiskImage.dmg*, buried inside Xcode. From the command line, you can attach the disk image and extract the debugserver to a local directory, as shown in Listing 6-1.

```
$ hdiutil attach /Applications/Xcode.app/Contents/Developer/Platforms/iPhoneOS.
    platform/DeviceSupport/7.1\ \(11D167\)/DeveloperDiskImage.dmg

Checksumming whole disk (Apple_HFS : 0)
..................................................................
    disk (Apple_HFS : 0): verified   CRC32 $D1221D77
    verified   CRC32 $B5681BED
    /dev/disk6            /Volumes/DeveloperDiskImage

$ cp /Volumes/DeveloperDiskImage/usr/bin/debugserver .
```

Listing 6-1: Extracting the debugserver from the Developer Disk Image

Once you've copied over the debugserver, you'll need to edit the entitlements of the binary. Normally, when Xcode itself uses the debugserver, it launches applications directly; you want to change its permissions to allow it to attach to arbitrary running programs on the device. First, generate a plist using the current entitlements of the binary, as follows:

```
$ codesign --display --entitlements entitlements.plist debugserver
```

This should result in an XML-formatted plist file with the following contents:

```
<?xml version="1.0" encoding="UTF-8"?>
<!DOCTYPE plist PUBLIC "-//Apple//DTD PLIST 1.0//EN" "http://www.apple.com/DTDs/
    PropertyList-1.0.dtd">
<plist version="1.0">
<dict>
        <key>com.apple.backboardd.debugapplications</key>
        <true/>
        <key>com.apple.backboardd.launchapplications</key>
```

```
        <true/>
        <key>com.apple.springboard.debugapplications</key>
        <true/>
        <key>run-unsigned-code</key>
        <true/>
        <key>seatbelt-profiles</key>
        <array>
                <string>debugserver</string>
        </array>
</dict>
</plist>
```

This file needs to be updated to include the get-task-allow and task_for_pid-allow entitlements and remove the seatbelt-profiles entitlement. Those updates will result in a plist like the following:

```
<!DOCTYPE plist PUBLIC "-//Apple//DTD PLIST 1.0//EN" "http://www.apple.com/DTDs/
    PropertyList-1.0.dtd">
<plist version="1.0">
<dict>
    <key>com.apple.springboard.debugapplications</key>
    <true/>
    <key>run-unsigned-code</key>
    <true/>
    <key>get-task-allow</key>
    <true/>
    <key>task_for_pid-allow</key>
    <true/>
</dict>
</plist>
```

After updating the *entitlements.plist* file, you'd use it to sign the application (thus overwriting the existing entitlements of the binary) and copy the debugserver to the device, as shown here:

```
$ codesign -s - --entitlements entitlements.plist -f debugserver
debugserver: replacing existing signature
$ scp debugserver root@de.vi.ce.ip:
```

Now you can finally debug the application. Ensure that the program you want to debug is currently running on the device and then launch the debugserver to attach to it, like this:

```
$ ssh root@de.vi.ce.ip
$ ./debugserver *:666 --attach=Snapchat

debugserver-310.2 for arm64.
```

```
Attaching to process Snapchat...
Listening to port 666 for a connection from *...
```

This example debugserver is now listening for a network connection from another machine running lldb. Next, on your local machine, you'd connect to the device as follows:

```
$ lldb
(lldb) platform select remote-ios
  Platform: remote-ios
 Connected: no
  SDK Path: "/Users/lx/Library/Developer/Xcode/iOS DeviceSupport/8.0 (12A4265u)"
 SDK Roots: [ 0] "/Applications/Xcode.app/Contents/Developer/Platforms/iPhoneOS.
     platform/DeviceSupport/4.2"
 SDK Roots: [ 1] "/Applications/Xcode.app/Contents/Developer/Platforms/iPhoneOS.
     platform/DeviceSupport/4.3"
 SDK Roots: [ 2] "/Applications/Xcode.app/Contents/Developer/Platforms/iPhoneOS.
     platform/DeviceSupport/5.0"
 SDK Roots: [ 3] "/Applications/Xcode.app/Contents/Developer/Platforms/iPhoneOS.
     platform/DeviceSupport/5.1"
 SDK Roots: [ 4] "/Applications/Xcode.app/Contents/Developer/Platforms/iPhoneOS.
     platform/DeviceSupport/6.0"
 SDK Roots: [ 5] "/Applications/Xcode.app/Contents/Developer/Platforms/iPhoneOS.
     platform/DeviceSupport/6.1"
 SDK Roots: [ 6] "/Applications/Xcode.app/Contents/Developer/Platforms/iPhoneOS.
     platform/DeviceSupport/7.0"
 SDK Roots: [ 7] "/Applications/Xcode.app/Contents/Developer/Platforms/iPhoneOS.
     platform/DeviceSupport/7.1 (11D167)"
 SDK Roots: [ 8] "/Users/lx/Library/Developer/Xcode/iOS DeviceSupport/5.0.1
     (9A405)"
 SDK Roots: [ 9] "/Users/lx/Library/Developer/Xcode/iOS DeviceSupport/6.0.1
     (10A523)"
 SDK Roots: [10] "/Users/lx/Library/Developer/Xcode/iOS DeviceSupport/7.0.4
     (11B554a)"
 SDK Roots: [11] "/Users/lx/Library/Developer/Xcode/iOS DeviceSupport/8.0
     (12A4265u)"
 SDK Roots: [12] "/Users/lx/Library/Developer/Xcode/iOS DeviceSupport/8.0
     (12A4297e)"

(lldb) process connect connect://de.vi.ce.ip:666
Process 2801 stopped
* thread #1: tid = 0x18b64b, 0x0000000192905cc0 libsystem_kernel.dylib`
    mach_msg_trap + 8, stop reason = signal SIGSTOP
    frame #0: 0x0000000192905cc0 libsystem_kernel.dylib`mach_msg_trap + 8
libsystem_kernel.dylib`mach_msg_trap + 8:
-> 0x192905cc0:  b        0x19290580c

libsystem_kernel.dylib`mach_msg_overwrite_trap:
   0x192905cc4:  .long   0x0000093a                  ; unknown opcode
```

```
0x192905cc8:  ldr    w16, 0x192905cd0              ; semaphore_signal_trap
0x192905ccc:  b      0x19290580c
```

In this example, the running program is now interrupted, and at this point, you'd be free to manipulate it with lldb on your local machine. To extract the decrypted program data, you'd next need to determine which part of the binary the encrypted segment resides in.

Note that you may find that a network connection is too unstable to complete the memory dump successfully. If this is the case, you can use the iproxy command included with usbmuxd to act as a proxy between your USB port and a TCP port, as follows:

```
$ brew install usbmuxd
$ iproxy 1234 1234 &
$ lldb
(lldb) process connect connect://127.0.0.1:1234
```

These commands connect to a network socket with lldb but actually go over the USB port.

Locating the Encrypted Segment

To locate the encrypted segment, you'll require odcctools and lldb. First, run otool -l *myBinary* and view the output in your favorite pager. You can do this either on the device or on your local machine. The copy included with OS X has a more modern version of otool that will provide cleaner output. Here's an example:

```
$ otool -fh Snapchat
Fat headers
fat_magic 0xcafebabe
nfat_arch 2
architecture 0
    cputype 12
    cpusubtype 9
    capabilities 0x0
    offset 16384
    size 9136464
    align 2^14 (16384)
architecture 1
    cputype 12
    cpusubtype 11
    capabilities 0x0
    offset 9158656
    size 9169312
    align 2^14 (16384)
Snapchat (architecture armv7:
Mach header
```

magic	cputype	cpusubtype	caps	filetype	ncmds	sizeofcmds	flags
0xfeedface	12	9	0x00	2	47	5316	0x00218085

Snapchat (architecture armv7s):
Mach header

magic	cputype	cpusubtype	caps	filetype	ncmds	sizeofcmds	flags
0xfeedface	12	11	0x00	2	47	5316	0x00218085

The Mach-O binary format allows for what are called *fat* files, which can contain the program compiled for multiple architectures at once (this is how OS X universal binaries work). To make reverse engineering easier, you need to work with the part of the binary that will be running on your target device; in my case, I have an iPhone 5s as a test device, so I want the armv7s architecture.

After determining the architecture, you have a couple of options. You could *thin* the binary to include only one architecture using the lipo(1) command (the thin flag specifies which architecture you're interested in), like this:

```
$ lipo -thin armv7 myBinary -output myBinary-thin
```

But for the purposes of this chapter, I'll show you how to work with a fat binary. First, you'd use otool to determine what the base address of the *text* segment of the binary is—this is where the actual executable instructions will be loaded into memory—as in Listing 6-2.

```
$ otool -arch armv7s -l Snapchat
Snapchat:
Load command 0
      cmd LC_SEGMENT
  cmdsize 56
  segname __PAGEZERO
   vmaddr 0x00000000
   vmsize 0x00004000
  fileoff 0
 filesize 0
  maxprot 0x00000000
 initprot 0x00000000
   nsects 0
    flags 0x0
Load command 1
      cmd LC_SEGMENT
  cmdsize 736
  segname __TEXT
   vmaddr 0x00004000
   vmsize 0x007a4000
  fileoff 0
 filesize 8011776
  maxprot 0x00000005
```

```
  initprot 0x00000005
    nsects 10
     flags 0x0
```

Listing 6-2: Finding the base address of the text segment

You can see here that the text segment starts at 0x00004000. Record this address because you'll need it in a bit. The next step is to determine the beginning and end of the encrypted part of the binary. You can do this with otool—note that you'll want to specify the `-arch armv7s` command (or whatever architecture you're using) to ensure that you're looking at the right section. The output should look like Listing 6-3.

```
$ otool -arch armv7s -l Snapchat
--snip--
Load command 9
      cmd LC_VERSION_MIN_IPHONEOS
  cmdsize 16
  version 5.0
      sdk 7.1
Load command 10
       cmd LC_UNIXTHREAD
   cmdsize 84
    flavor ARM_THREAD_STATE
     count ARM_THREAD_STATE_COUNT
           r0   0x00000000 r1    0x00000000 r2  0x00000000 r3  0x00000000
           r4   0x00000000 r5    0x00000000 r6  0x00000000 r7  0x00000000
           r8   0x00000000 r9    0x00000000 r10 0x00000000 r11 0x00000000
           r12 0x00000000 sp    0x00000000 lr  0x00000000 pc  0x0000a300
          cpsr 0x00000000
Load command 11
       cmd LC_ENCRYPTION_INFO
   cmdsize 20
cryptoff 16384
cryptsize 7995392
cryptid 1
```

Listing 6-3: otool displaying a binary's load commands

The values of interest here are `cryptoff` and `cryptsize` (`cryptid` simply indicates this is an encrypted binary).[5] These indicate the address where the encrypted segment of the application begins and the size of the segment, respectively. The range between those two numbers will help you when dumping memory. These values are in hexadecimal, though—a quick way to obtain the hex values is to execute the following in the Terminal:

---

5. *https://developer.apple.com/library/mac/#documentation/DeveloperTools/Conceptual/MachORuntime/Reference/reference.html*

```
$ printf '%x\n' 16384
4000
$ printf '%x\n' 7995392
7a0000
```

In this case, the numbers are 0x00004000 and 0x007a0000. Write these down, too. Now, back in Listing 6-2, it was determined that the text segment in the binary starts at 0x00004000. However, the text segment probably won't end up there when the program is actually run because ASLR moves portions of memory around at random.[6] So check to see where the text segment actually got loaded using lldb's image list command, as follows:

```
(lldb) image list
[   0] E3BB2396-1EF8-3EA7-BC1D-98F736A0370F 0x000b2000 /var/mobile/Applications/
       CCAC51DD-48DB-4798-9D1B-94C5C700191F/Snapchat.app/Snapchat
       (0x00000000000b2000)
[   1] F49F2879-0AA0-36C0-8E55-73071A7E2870 0x2db90000 /Users/lx/Library/Developer/
       Xcode/iOS DeviceSupport/7.0.4 (11B554a)/Symbols/System/Library/Frameworks/
       AudioToolbox.framework/AudioToolbox
[   2] 763DDFFB-38AF-3444-B745-01DDE37A5949 0x388ac000 /Users/lx/Library/Developer/
       Xcode/iOS DeviceSupport/7.0.4 (11B554a)/Symbols/usr/lib/libresolv.9.dylib
[   3] 18B3A243-F792-3C39-951C-97AB416ED3E6 0x37fb0000 /Users/lx/Library/Developer/
       Xcode/iOS DeviceSupport/7.0.4 (11B554a)/Symbols/usr/lib/libc++.1.dylib
[   4] BC1A8B9C-9F5D-3B9D-B79E-345D4C3A361A 0x2e7a2000 /Users/lx/Library/Developer/
       Xcode/iOS DeviceSupport/7.0.4 (11B554a)/Symbols/System/Library/Frameworks/
       CoreLocation.framework/CoreLocation
[   5] CC733C2C-249E-3161-A9AF-19A44AEB1577 0x2d8c2000 /Users/lx/Library/Developer/
       Xcode/iOS DeviceSupport/7.0.4 (11B554a)/Symbols/System/Library/Frameworks/
       AddressBook.framework/AddressBook
```

You can see that the text segment landed at 0x000b2000. With that address in hand, you're finally ready to extract the executable part of the binary.

Dumping Application Memory

Let's look at a bit of math to figure out the final offsets. The first step is to add the base address to the value of cryptoff; in this case, both were 0x00004000, so the starting number would be 0x00008000. The ending number would be the starting number plus the value of cryptsize, which is at 0x007a0000 in this example. These particular numbers are pretty easy to add in your head, but if you get offsets you can't figure out easily, you can just use Python to calculate it for you, as shown in Listing 6-4.

---

6. Unless you disable PIE. You can do this with the removePIE tool; see *https://github.com/ peterfillmore/removePIE/*.

```
$ python
Python 2.7.10 (default, Dec 14 2015, 19:46:27)
[GCC 4.2.1 Compatible Apple LLVM 6.0 (clang-600.0.39)] on darwin
Type "help", "copyright", "credits" or "license" for more information.
>>>
>>> hex(0x00008000 + 0x007a0000)
'0x7a8000'
```

Listing 6-4: Adding the starting number and the hexadecimal value of cryptsize

Now this example is seriously almost done, I promise. From here, you'd just plug your numbers in to the following lldb command:

```
(lldb) memory read --force --outfile /tmp/mem.bin --binary 0x00008000 0x007a8000
8011776 bytes written to '/private/tmp/mem.bin'
```

This won't give you a full, working binary, of course—just a memory dump. The image lacks the Mach-O header metadata. To fix this, you'd need to transplant the memory dump into a valid binary, and to that end, you'd first make a copy of the original binary and use scp to copy it to your development machine.

Then, you'd copy the contents of the unencrypted memory dump into the donor binary, replacing the encrypted segment. You can use dd for this, specifying the seek parameter where it should start writing your data. The seek parameter should be the value of vmaddr added to cryptoff, which is 0x8000 in this case. Here's how this example's dd command would look:

```
$ dd bs=1 seek=0x8000 conv=notrunc if=/tmp/mem.bin of=Snapchat-decrypted
```

Next, you'd have to change the donor binary's cryptid value to 0, indicating an unencrypted binary. There are several ways to do this. You can use MachOView[7] (see Figure 6-2), which provides an easy interface for examining and changing Mach-O binaries, or you can use a hex editor of your choice. If you're using a hex editor, I find it easiest to first find the LC_ENCRYPTION_INFO command by searching for 2100 0000 1400 0000.[8] The next 16 numbers will be the offset and size, followed by 0100 0000. That byte is the cryptid; change it to 0000 0000.

Once you've disabled the cryptid flag, you'd need to copy the modified binary back to the device. With the modified binary in place, you can verify the change using vbindiff, which is available in Homebrew. Output from vbindiff should appear as shown in Listing 6-5.

---

7. *http://sourceforge.net/projects/machoview/*

8. This is how it appears in xxd(1), which is what I usually use for quick-and-dirty editing. Your editor may vary. If in doubt, check with MachOView first and then develop whatever scripts you may require.

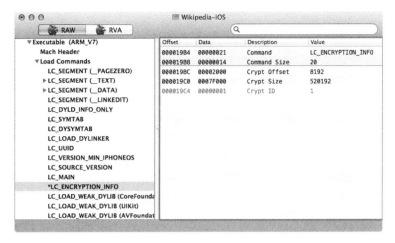

Figure 6-2: The encrypted flag with MachOView

```
Snapchat
0000 0A80: 00 00 00 00 00 00 00 00   00 00 00 00 00 00 00 00   ........ ........
0000 0A90: 00 00 00 00 00 00 00 00   00 00 00 00 00 00 00 00   ........ ........
0000 0AA0: 00 00 00 00 70 39 00 00   00 00 00 00 21 00 00 00   ....p9.. ....!...
❶ 0000 0AB0: 14 00 00 00 00 20 00 00   00 C0 26 01 01 00 00 00   ..... .. ..&.....
0000 0AC0: 0C 00 00 00 34 00 00 00   18 00 00 00 02 00 00 00   ....4... ........
0000 0AD0: 00 00 01 00 00 00 01 00   2F 75 73 72 2F 6C 69 62   ........ /usr/lib
0000 0AE0: 2F 6C 69 62 6C 6F 63 6B   64 6F 77 6E 2E 64 79 6C   /liblock down.dyl
Snapchat-decrypted
0000 0A80: 00 00 00 00 00 00 00 00   00 00 00 00 00 00 00 00   ........ ........
0000 0A90: 00 00 00 00 00 00 00 00   00 00 00 00 00 00 00 00   ........ ........
0000 0AA0: 00 00 00 00 70 39 00 00   00 00 00 00 21 00 00 00   ....p9.. ....!...
❷ 0000 0AB0: 14 00 00 00 00 20 00 00   00 C0 26 01 00 00 00 00   ..... .. ..&.....
0000 0AC0: 0C 00 00 00 34 00 00 00   18 00 00 00 02 00 00 00   ....4... ........
0000 0AD0: 00 00 01 00 00 00 01 00   2F 75 73 72 2F 6C 69 62   ........ /usr/lib
0000 0AE0: 2F 6C 69 62 6C 6F 63 6B   64 6F 77 6E 2E 64 79 6C   /liblock down.dyl
```

Listing 6-5: Verifying the changed `cryptid` value with `vbindiff`

The lines at ❶ and ❷ show the `cryptid` bit (in bold) enabled and disabled, respectively. Now, if all has gone well, you'd be ready to start dissecting the binary in earnest.

Reverse Engineering from Decrypted Binaries

Because of the rather transparent structure of the Mach-O binary format, basic reverse engineering on iOS is a fairly trivial task—at least once you've managed to obtain a decrypted binary. Several tools can help you understand class definitions, examine assembly instructions, and give details on how the binary was built. The most useful and easily obtainable ones are otool and class-dump. You'll also take a look at Cycript and Hopper as tools for reversing particularly stubborn applications.

Inspecting Binaries with otool

otool has long been part of the base OS X toolkit for inspecting Mach-O binaries. Its current incarnation supports both ARM and amd64 architectures and can optionally use llvm to disassemble binaries. To get a basic look at a program's internals, you can use otool -oV to view the data segment, as shown in Listing 6-6.

```
$ otool -oV MobileMail

MobileMail:
Contents of (__DATA,__objc_classlist) section
000c2870 0xd7be8
          isa 0xd7bd4
   superclass 0x0
        cache 0x0
       vtable 0x0
         data 0xc303c (struct class_ro_t *)
                flags 0x0
        instanceStart 80
         instanceSize 232
           ivarLayout 0x0
                 name 0xb48ac MailAppController
          baseMethods 0xc3064 (struct method_list_t *)
              entsize 12
                count 122
                 name 0xa048e toolbarFixedSpaceItem
                types 0xb5bb0 @8@0:4
                  imp 0x40c69
                 name 0xa04a4 sidebarQuasiSelectTintColor
                types 0xb5bb0 @8@0:4
                  imp 0x40ccd
                 name 0xa04c0 sidebarMultiselectTintColor
                types 0xb5bb0 @8@0:4
                  imp 0x40d75
                 name 0xa04dc sidebarTintColor
                types 0xb5bb0 @8@0:4
                  imp 0x130f5
                 name 0xa04ed updateStyleOfToolbarActivityIndicatorView:
     inView:
                types 0xb5c34 v16@0:4@8@12
                  imp 0x18d69
```

Listing 6-6: otool displaying the contents of the __OBJC segment

This gives you a view of class and method names, as well as information about ivars, provided these are implemented in Objective-C rather than straight C++. To view the text segment of a program, you can use otool -tVq. The -q indicates that you want to use llvm as the disassembler rather than otool's built-in disassembler, which is noted by -Q. The differences in output are few, but llvm seems best suited for the task, given that it likely assembled the binary in the first place. It also provides slightly more readable output. Listing 6-7 shows some example output of otool -tVq.

```
MobileMail:
(__TEXT,__text) section
00003584        0000        movs    r0, r0
00003586        e59d        b       0x30c4
00003588        1004        asrs    r4, r0, #32

--snip--

000035ca        447a        add     r2, pc
000035cc        6801        ldr     r1, [r0]
000035ce        6810        ldr     r0, [r2]
000035d0        f0beecf0    blx     0xc1fb4 @ symbol stub for: _objc_msgSend
000035d4        f2417128    movw    r1, #5928
000035d8        f2c0010d    movt    r1, #13
000035dc        4479        add     r1, pc
000035de        6809        ldr     r1, [r1]
000035e0        f0beece8    blx     0xc1fb4 @ symbol stub for: _objc_msgSend
000035e4        4606        mov     r6, r0
```

Listing 6-7: otool's disassembly output

Here, you see the actual disassembly of methods, as well as some basic symbol information. To get a dump of all the symbols, use otool -IV, as shown in Listing 6-8.

```
$ otool -IV MobileMail

MobileMail:
Indirect symbols for (__TEXT,__symbolstub1) 241 entries
address     index name
0x000c1c30      3 _ABAddressBookFindPersonMatchingEmailAddress
0x000c1c34      4 _ABAddressBookRevert
0x000c1c38      5 _ABPersonCopyImageDataAndCropRect
0x000c1c3c      7 _CFAbsoluteTimeGetCurrent
0x000c1c40      8 _CFAbsoluteTimeGetGregorianDate
0x000c1c44      9 _CFArrayAppendValue
0x000c1c48     10 _CFArrayCreateMutable
0x000c1c4c     11 _CFArrayGetCount
```

```
0x000c1c50    12 _CFArrayGetFirstIndexOfValue
0x000c1c54    13 _CFArrayGetValueAtIndex
0x000c1c58    14 _CFArrayRemoveValueAtIndex
0x000c1c5c    15 _CFArraySortValues
0x000c1c60    16 _CFDateFormatterCopyProperty
0x000c1c64    17 _CFDateFormatterCreate
```

Listing 6-8: Inspecting symbols with otool

Obtaining Class Information with class-dump

The class-dump[9] tool is used to extract class information from Objective-C 2.0 binaries. The resulting output is essentially the equivalent of the header files of a given binary. This can give excellent insight into the design and structure of a program, making class-dump an invaluable tool for reverse engineering. The original class-dump by Steve Nygard runs only on OS X but recognizes the armv7 architecture, so you can copy files over to your desktop for analysis. There is also a modified version, class-dump-z,[10] that can run on Linux and iOS. As of this writing, class-dump appears to be more up-to-date and functional, so I recommend sticking with it.

You can test class-dump against any unencrypted iOS binary. The quickest way to get a feel for it is to copy over one of the built-in Apple apps in */Applications* and run class-dump on the binary, as shown in Listing 6-9.

```
$ class-dump MobileMail

--snip--
@interface MessageHeaderHeader : _AAAccountConfigChangedNotification <
    MessageHeaderAddressBookClient, UIActionSheetDelegate>
{
    MailMessage *_lastMessage;
    id <MessageHeaderDelegate> _delegate;
    UIWebBrowserView *_subjectWebView;
    DOMHTMLElement *_subjectTextElement;
    UILabel *_dateLabel;
    unsigned int _markedAsUnread:1;
    unsigned int _markedAsFlagged:1;
    unsigned int _isOutgoing:1;
    UIImageView *_unreadIndicator;
    UIImageView *_flaggedIndicator;
    WorkingPushButton *_markButton;
    id _markUnreadTarget;
```

---

9. *http://stevenygard.com/projects/class-dump/*

10. *http://code.google.com/p/networkpx/wiki/class_dump_z*

```
    SEL _markUnreadAction;
    ABPersonIconImageView *_personIconImageView;
    SeparatorLayer *_bottomSeparator;
    SeparatorLayer *_topSeparator;
    float _horizontalInset;
    unsigned int _allowUnreadStateToBeShown:1;
}


- (id)initWithFrame:(struct CGRect)fp8;
- (void)dealloc;
```

Listing 6-9: class-dump showing the interface details of MobileMail

Delightful, no? Once you have a decrypted binary, most Objective-C applications become transparent pretty quickly.

Extracting Data from Running Programs with Cycript

If you don't want to go through the hassle of decrypting a binary to get information about its internals, you can use Cycript[11] to extract some of this information from a running executable. There are many tricks to interact with running applications using Cycript, but you'll probably be most interested in using *weak_classdump.cy*[12] to approximate the functionality of class-dump. With the Contacts application running, you can extract class-dump information thusly:

```
$ curl -OL https://raw.github.com/limneos/weak_classdump/master/
    weak_classdump.cy
$ cycript -p Contacts weak_classdump.cy
'Added weak_classdump to "Contacts" (3229)'
$ cycript -p Contacts
cy# weak_classdump_bundle([NSBundle mainBundle],"/tmp/contactsbundle")
"Dumping bundle... Check syslog. Will play lock sound when done."
```

This will write out header files for each class into the */tmp/contactsbundle* directory.

Note that in order to securely fetch things with cURL, you'll need to install a CA certificate bundle on the device. If you use MacPorts and have cURL installed locally, do this:

```
$ scp /opt/local/share/curl/curl-ca-bundle.crt \
    root@de.vi.c.e:/etc/ssl/certificates/ca-certificates.crt
```

---

11. *http://www.cycript.org/*

12. *https://github.com/limneos/weak_classdump/*

Or if you use Homebrew and have the OpenSSL formula installed, you can use this command:

```
$ scp /usr/local/etc/openssl/cert.pem \
    root@de.vi.c.e:/etc/ssl/certificates/ca-certificates.crt}
```

Disassembly with Hopper

There will likely be some situations where you need to get a closer view of a program's actual logic, in the absence of source code. While IDA Pro[13] is useful for this, it's rather expensive. I usually use Hopper[14] for disassembling, decompiling, and making flow graphs during black-box testing. While assembly language and decompiling are somewhat outside the scope of this book, let's take a quick look at what Hopper can show you about a program's logic. Looking at a basic password manager in Hopper (Figure 6-3), you will find a method called storeSavedKeyFor:, which looks promising.

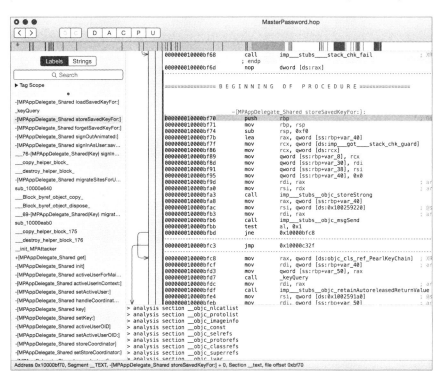

Figure 6-3: The disassembly of the storeSavedKeyFor: function

13. *https://www.hex-rays.com/products/ida/*

14. *http://www.hopperapp.com/*

If you call the decompiler (the if(b) button) on this particular section of code, Hopper will generate pseudocode to give you an idea of actual program flow, as shown in Figure 6-4.

```
function -[MPAppDelegate_Shared storeSavedKeyFor:] {
    var_216 = **__stack_chk_guard;
    var_208 = rdi;
    var_200 = rsi;
    var_192 = 0x0;
    objc_storeStrong(&var_192, rdx);
    if ((LOBYTE([var_192 saveKey]) & 0x1) != 0x0) {
            var_144 = *objc_cls_ref_PearlKeyChain;
            rax = [_keyQuery(var_192) retain];
            var_136 = rax;
            var_184 = [[var_144 dataOfItemForQuery:rax] retain];
            [var_136 release];
            var_128 = var_184;
            rax = [[var_208 key] retain];
            var_120 = rax;
            rax = [[rax keyData] retain];
            var_112 = rax;
            var_111 = LOBYTE(LOBYTE([var_128 isEqualToData:rax]) ^ 0x1);
            [var_112 release];
            [var_120 release];
            if ((LOBYTE(var_111) & 0x1) != 0x0) {
                    var_96 = [[PearlLogger get] retain];
                    rax = basename("/Users/lx/git/MasterPassword/
MasterPassword/ObjC/MPAppDelegate_Key.m");
                    var_88 = *0x100203170;
                    var_80 = rax;
                    rax = [[var_192 userID] retain];
                    var_72 = *0x100203178;
                    var_64 = @"Saving key in keychain for: %@";
                    var_56 = rax;
                    [[[var_96 inFile:rdx atLine:rcx inf:r8] retain]
release];
                    [var_56 release];
                    [var_96 release];
                    var_48 = *objc_cls_ref_PearlKeyChain;
                    rax = [_keyQuery(var_192) retain];
                    var_152 = **kSecValueData;
                    var_40 = rax;
                    rax = [[var_208 key] retain];
                    var_32 = rax;
                    rax = [[rax keyData] retain];
                    var_168 = rax;
                    var_160 = **kSecAttrAccessible;
                    var_176 =
**kSecAttrAccessibleWhenUnlockedThisDeviceOnly;
                    var_24 = *0x100203108;
                    var_16 = rax;
                    rax = [[NSDictionary dictionaryWithObjects:rdx
forKeys:rcx count:r8] retain];
                    var_8 = rax;
                    var_4 = LODWORD([var_48 addOrUpdateItemForQuery:var_40
withAttributes:rax]);
                    [var_8 release];
                    [var_16 release];
                    [var_32 release];
                    [var_40 release];
            }
            objc_storeStrong(&var_184, 0x0);
    }
    objc_storeStrong(&var_192, 0x0);
```

Figure 6-4: Code generated by the decompiler

Notice that the PearlLogger class is being instantiated, and there's a reference to the username for which the current item is being stored. var_64 shows that this username is getting passed to the logging function, probably to the NSLog facility—this is bad, for reasons I'll explain further in Chapter 10. However, you can also see that the item is being stored in the Keychain with a restrictive protection attribute (kSecAttrAccessibleWhen-UnlockedThisDeviceOnly, further detailed in Chapter 13), which is a point in the program's favor.

Assembly language and decompilation are broad areas, but Hopper gives you a great way to get started with reverse engineering via assembly for a fairly low price. If you'd like to get started developing your skills reading ARM assembly, check out Ray Wenderlich's tutorial: *http://www.raywenderlich.com/37181/ios-assembly-tutorial/*.

Defeating Certificate Pinning

Certificate pinning aims to prevent a rogue CA from signing a fake (but valid-looking) certificate for your site, with the purpose of intercepting communications between your network endpoint and the application. This is quite a good idea (and I'll discuss how to implement it in Chapter 7), but it does of course make black-box testing slightly more difficult.

My colleagues and I ran into this problem frequently enough that we wrote a tool to help us with it: the iOS SSL Killswitch.[15] The Killswitch tool hooks requests going through the URL loading system to prevent the validation of any SSL certificates, ensuring that you can run any black-box application through your proxy regardless of whether it uses certificate pinning.

To install the Killswitch tool, copy the precompiled *.deb* file to your device and install it with the dpkg tool.

```
# scp ios-ssl-kill-switch.deb root@192.168.1.107
# ssh root@192.168.1.107
(and then, on the test device)
# dpkg -i ios-ssl-kill-switch.deb
# killall -HUP SpringBoard
```

You should then find iOS SSL Killswitch in your Settings application (see Figure 6-5), where you can toggle it on and off.

Figure 6-5: Enabling the SSL Killswitch tool from within the Settings application

---

15. *https://github.com/iSECPartners/ios-ssl-kill-switch/*

Hooking with Cydia Substrate

On jailbroken devices (which you'll be performing your black-box testing on), you can use Cydia Substrate[16] (formerly known as Mobile Substrate) to modify the behavior of the base system to give you additional information on your application's activity or change application behavior. Your goals may be to disable certain security or validation mechanisms (like the iOS SSL Killswitch does) or to simply notify you when certain APIs are used, along with the arguments passed to them. Cydia Substrate hooks are referred to as *tweaks.*

The most user-friendly way to get started with developing Cydia Substrate tweaks is to use the Theos toolkit.[17] To create a new tweak, use the *nic.pl* script included with Theos. Note that Theos is by default oriented toward tweaking the behavior of the SpringBoard application in order to customize user interface elements. For the purposes described in this book, though, you'll want to affect all applications, so you'd specify a Bundle filter of com.apple.UIKit. This filter will configure Mobile/Cydia Substrate to load your tweak in any application that links to the UIKit framework (that is, applications displaying a user interface) but not other programs like system daemons or command line tools.

First, you need to acquire the Link Identity Editor, ldid,[18] which Theos uses to generate the signature and entitlements for a tweak. Here's how to get ldid:

```
$ git clone git://git.saurik.com/ldid.git

$ cd ldid

$ git submodule update --init

$ ./make.sh

$ sudo cp ./ldid /usr/local/bin
```

You can then clone the Theos repo and proceed to generate a tweak template, as follows:

```
$ git clone git://github.com/DHowett/theos.git ~/git/theos

$ cd /tmp && ~/git/theos/bin/nic.pl
NIC 2.0 - New Instance Creator
------------------------------
  [1.] iphone/application
```

---

16. *http://iphonedevwiki.net/index.php/MobileSubstrate*

17. *http://iphonedevwiki.net/index.php/Theos/Getting_Started*

18. *http://gitweb.saurik.com/ldid.git*

```
  [2.] iphone/library
  [3.] iphone/preference_bundle
  [4.] iphone/tool
  [5.] iphone/tweak
Choose a Template (required): 5
Project Name (required): MyTweak
Package Name [com.yourcompany.mytweak]:
Author/Maintainer Name [dthiel]:
[iphone/tweak] MobileSubstrate Bundle filter [com.apple.springboard]: com.apple.
    UIKit
Instantiating iphone/tweak in mytweak/...
Done.
```

This will create a *Tweak.xm* file, with all of its contents commented out by default. Stubs are included for hooking either class methods or instance methods, with or without arguments.

The simplest type of hook you can write is one that just logs method calls and arguments; here's an example that hooks two class methods of UIPasteboard:

```
%hook UIPasteboard

+ (UIPasteboard *)pasteboardWithName:(NSString *)pasteboardName create:(BOOL)create
{
      %log;
      return %orig;
}

+ (UIPasteboard *)generalPasteboard
{
      %log;
      return %orig;
}

%end
```

This code snippet uses Logos[19] directives such as %hook and %log. Logos is a component of Theos designed to allow method-hooking code to be written easily. However, it is possible to write a tweak with the same functionality using only C instead.

You'll want to provide the full method signature as well, which you can obtain either from API documentation or from framework header files. Once you've customized your tweak to your satisfaction, you can build it using the Makefile provided by *nic.pl*.

---

19. *http://iphonedevwiki.net/index.php/Logos*

To build a Debian package suitable for installation onto a jailbroken device, you'll also need to install the dpkg tool. You can do this either with the MacPorts[20] port command or with Homebrew's[21] brew command. This example uses port:

```
$ sudo port install dpkg
    --snip--
$ make
Bootstrapping CydiaSubstrate...
 Compiling iPhoneOS CydiaSubstrate stub... default target?
 Compiling native CydiaSubstrate stub...
 Generating substrate.h header...
Making all for tweak MyTweak...
 Preprocessing Tweak.xm...
 Compiling Tweak.xm...
 Linking tweak MyTweak...
 Stripping MyTweak...
 Signing MyTweak...
$ make package
Making all for tweak MyTweak...
make[2]: Nothing to be done for `internal-library-compile'.
Making stage for tweak MyTweak...
dpkg-deb: building package `com.yourcompany.mytweak' in `./com.yourcompany.
    mytweak_0.0.1-1_iphoneos-arm.deb'.
```

Running these commands should result in a package that can be installed on your iOS device. First, you'd use the scp command to copy the file over to the device and load it manually. After that, you could simply use dpkg -i from the command line (as shown in the following code) or set up your own Cydia repository.[22]

```
$ dpkg -i com.yourcompany.mytweak_0.0.1-1_iphoneos-arm.deb
Selecting previously deselected package com.yourcompany.mytweak.
(Reading database ... 3551 files and directories currently installed.)
Unpacking com.yourcompany.mytweak (from com.yourcompany.mytweak_0.0.1-1_iphoneos-
    arm.deb) ...
Setting up com.yourcompany.mytweak (0.0.1-1) ..
```

When this finishes, you can either manage the package further with the dpkg command (removing it with dpkg -P) or manage it via Cydia, as shown in Figure 6-6.

---

20. *http://www.macports.org/*

21. *http://brew.sh/*

22. *http://www.saurik.com/id/7/*

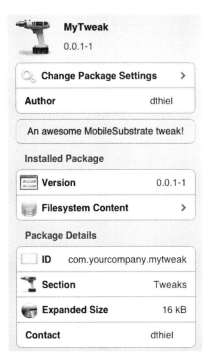

Figure 6-6: Your very own tweak in the Cydia management interface

After a tweak is installed, if you examine the system log, you'll see the Cydia Substrate dynamic library being loaded upon launch of all applications. You'll also see the hooked method calls being logged by the tweak. Here's an example log:

```
May  2 14:22:08 my-iPad Maps~ipad[249]: MS:Notice: Loading: /Library/
    MobileSubstrate/DynamicLibraries/MyTweak.dylib
May  2 14:22:38 lxs-iPad Maps~ipad[249]: +[<UIPasteboard: 0x3ef05408>
    generalPasteboard]
```

There are, of course, many other things you can do with tweaks besides logging; see the *Tweak.xm* file of the iOS SSL Killswitch tool for an example of modifying method behavior, along with your own preference toggle.[23]

Automating Hooking with Introspy

While tweaks are useful for one-off hooking scenarios, my colleagues Alban Diquet and Tom Daniels have used the Cydia Substrate framework to make a tool called Introspy[24] that can help automate the hooking process for

---

23. *https://github.com/iSECPartners/ios-ssl-kill-switch/blob/master/Tweak.xm*

24. *https://github.com/iSECPartners/Introspy-iOS/*

black-box testing without having to dig too deep in to the guts of iOS or Cydia Substratey. Introspy uses the Cydia Substrate framework directly (rather than via Theos) to hook security-sensitive method calls, logging their arguments and return values in a format that can subsequently be used to generate reports. To install Introspy, download the latest precompiled *.deb* package from *https://github.com/iSECPartners/Introspy-iOS/releases/*, copy it to your device, and enter the command dpkg -i *filename* on the device to add the package.

Once installed, respring the device using the following:

```
$ killall -HUP SpringBoard
```

Do the same for any application that you want to test, if it's already running. You can now tell Introspy what applications you want to hook, along with which API calls you'd like to record (see Figure 6-7). Once your testing is complete, a SQLite database file will be deposited in */var/mobile* if you're testing Apple built-in or Cydia applications, or in */User/Applications/<AppID>* if you're testing an application that came from the App Store.

Figure 6-7: The Introspy settings screen. You can select which applications are profiled on the Apps tab.

To analyze this database, you'll want to use the Introspy Analyzer,[25] which will generate HTML reports of Introspy's findings (see Figure 6-8).

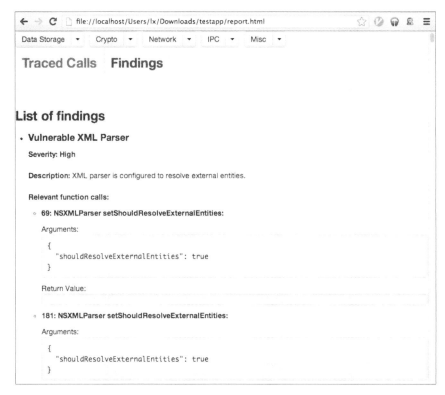

Figure 6-8: The Introspy HTML report output, showing a list of findings that match the specified signatures

If you copy this database onto your test machine, you can create a report on the called APIs using *introspy.py*, as follows:

```
$ python ./introspy.py --outdir report mydatabase.db
```

Newer versions of Introspy also allow automatic copying and parsing of the database, by specifying the IP address of the device.

```
$ python ./introspy.py -p ios -o outputdir -f device.ip.address
```

25. *https://github.com/iSECPartners/Introspy-Analyzer/*

Running Introspy will evaluate the calls against a signature database of potentially problematic APIs, helping you track down potential areas of interest. To cut down on noise, you can filter out specific API categories or signature types with the `--group` and `--sub-group` flags. With Introspy installed, enter `introspy.py --help` at the command line for details.

Closing Thoughts

While black-box testing poses some challenges, the development community has gone a long way to making it feasible, and some elements of black-box testing will help you regardless of whether you have source code. You will now turn your primary attention back to white-box testing; in Chapter 7, I'll guide you through some of the most security-sensitive APIs in iOS, including IPC mechanisms, cryptographic facilities, and the myriad ways in which data can leak from applications unintentionally.

PART III

SECURITY QUIRKS OF THE COCOA API

7

iOS NETWORKING

Almost all applications use one or more of three
iOS network APIs. In order of abstraction, these are
the URL loading system, the Foundation `NSStream`
API, and the Core Foundation `CFStream` API. The URL
loading system is used for fetching and manipulating
data, such as network resources or files, via URLs. The
`NSStream` and `CFStream` classes are slightly lower-level methods to deal with
network connections, without going quite so low as the socket level. These
classes are used for non-HTTP-based communications, or where you need
more direct control over network behavior.

In this chapter, I'll discuss iOS networking in detail, starting from the
high-level APIs. For most purposes, apps can stick with the higher-level APIs,
but there are some cases where you can't quite bend those APIs to your will.
With lower-level APIs, however, there are more pitfalls to consider.

Using the iOS URL Loading System

The URL loading system can handle most network tasks an app will need to perform. The primary method of interacting with the URL API is by constructing an `NSURLRequest` object and using it to instantiate an `NSURLConnection` object, along with a delegate that will receive the connection's response. When the response is fully received, the delegate will be sent a `connection:didReceiveResponse` message, with an `NSURLResponse` object as the supplied parameter.[1]

But not everyone uses the powers of the URL loading system properly, so in this section, I'll first show you how to spot an app that bypasses Transport Layer Security. Then, you'll learn how to authenticate endpoints through certificates, avoid the dangers of open redirects, and implement certificate pinning to limit how many certificates your app trusts.

Using Transport Layer Security Correctly

Transport Layer Security (TLS), the modern specification supplanting SSL, is crucial to the security of almost any networked application. When used correctly, TLS both keeps the data transmitted over a connection confidential and authenticates the remote endpoint, ensuring that the certificate presented is signed by a trusted certificate authority. By default, iOS does the Right Thing™ and refuses to connect to any endpoint with an untrusted or invalid certificate. But all too frequently, in applications of all kinds, mobile and otherwise, developers explicitly disable TLS/SSL endpoint validation, allowing the application's traffic to be intercepted by network attackers.

In iOS, TLS can be disabled a number of ways. In the past, developers would often use the undocumented `setAllowsAnyHTTPSCertificate` private class method of `NSURLRequest` to easily disable verification. Apple fairly quickly started rejecting applications that used this method, as it tends to do with apps that use private APIs. There are, however, still obfuscation methods that may allow the use of this API to slip past the approval process, so check codebases to ensure that the method isn't just called by another name.

There's an even more disastrous way to bypass TLS validation. It will also (probably) get your app rejected in this day and age, but it illustrates an important point about categories. I once had a client that licensed what should have been a fairly simple piece of third-party code and included it in their product. Despite handling TLS correctly everywhere else in the project, their updated version of the third-party code did not validate any TLS connections. Apparently, the third-party vendor had implemented a category of `NSURLRequest`, using the `allowsAnyHTTPSCertificateForHost` method to avoid validation. The category contained only the directive `return YES;`, causing all `NSURLRequests` to silently ignore bad certificates. The moral? Test things, and don't make assumptions! Also, you have to audit third-party code

---

1. *https://developer.apple.com/DOCUMENTATION/Cocoa/Conceptual/URLLoadingSystem/URLLoadingSystem.pdf*

along with the rest of your codebase. Mistakes might not be your fault, but nobody is likely to care about that.

NOTE *Thankfully, it's much more difficult to make accidental TLS-disabling mistakes in iOS 9, as it by default does not allow applications to make non-TLS connections. Instead, developers are required to put a specific exception in the app's* Info.plist *for URLs to be accessed over plaintext HTTP. However, this won't solve cases of willful disabling of TLS protections.*

Now, there is actually an official API to bypass TLS verification. You can use a delegate of NSURLConnection with the NSURLConnectionDelegate protocol.[2] The delegate must implement the willSendRequestForAuthenticationChallenge method, which can then call the continueWithoutCredentialForAuthentication-Challenge method. This is the current, up-to-date method; you may also see older code that uses connection:canAuthenticateAgainstProtectionSpace: or connection:didReceiveAuthenticationChallenge:. Listing 7-1 shows an example of how you might see this done in the wild.

```
- (void)connection:(NSURLConnection *)connection
    willSendRequestForAuthenticationChallenge:(NSURLAuthenticationChallenge *)
    challenge {
    NSURLProtectionSpace *space = [challenge protectionSpace];
    if([[space authenticationMethod] isEqualToString:
    NSURLAuthenticationMethodServerTrust]) {
        NSURLCredential *cred = [NSURLCredential credentialForTrust:
    [space serverTrust]];
        [[challenge sender] useCredential:cred forAuthenticationChallenge:
    challenge];
    }
}
```

Listing 7-1: Sending a dummy NSURLCredential *in response to the challenge*

This code looks rather benign, especially since it uses the words *protection*, *credential*, *authentication*, and *trust* all over the place. What it actually does is bypass verification of the TLS endpoint, leaving the connection susceptible to interception.

Of course, I'm not encouraging you to actually *do* anything to bypass TLS verification in your app. You shouldn't, and you're a bad person if you do. These examples just show the pattern that you may see in code that you have to examine. These patterns can be difficult to spot and understand, but if you see code that bypasses TLS verification, be sure to change it.

_____

2. *https://developer.apple.com/library/mac/#documentation/Foundation/Reference/ NSURLConnectionDelegate_Protocol*

Basic Authentication with NSURLConnection

HTTP basic authentication isn't a particularly robust authentication mechanism. It doesn't support session management or password management, and therefore, the user can't log out or change their password without using a separate application. But for some tasks, such as authenticating to APIs, these issues are less important, and you still might run across this mechanism in an app's codebase—or be required to implement it yourself.

You can implement HTTP basic authentication using either NSURLSession or NSURLConnection, but there are a couple of pitfalls that you'll want to be aware of, whether you're writing an app or examining someone else's code.

The simplest implementation uses the willSendRequestForAuthentication-Challenge delegate method of NSURLConnection:

```
- (void)connection:(NSURLConnection *)connection
    willSendRequestForAuthenticationChallenge:(NSURLAuthenticationChallenge *)
    challenge {
  NSString *user = @"user";
  NSString *pass = @"pass";


  if ([[challenge protectionSpace] receivesCredentialSecurely] == YES &&
      [[[challenge protectionSpace] host] isEqualToString:@"myhost.com"]) {

  NSURLCredential *credential = [NSURLCredential credentialWithUser:user password
    :pass persistence:NSURLCredentialPersistenceForSession];

  [[challenge sender] useCredential:credential
        forAuthenticationChallenge:challenge];
  }
}
```

The delegate is first passed an NSURLAuthenticationChallenge object. It then creates a credential with a username and password, which can be either provided by the user or pulled from the Keychain. Finally, the sender of the challenge is passed the credential and challenge in return.

There are two potential problems to pay attention to when implementing HTTP basic authentication in this way. First, avoid storing the username and password within either the source code or the shared preferences. You can use the NSURLCredentialStorage API to store user-supplied credentials in the Keychain automatically, using sharedCredentialStorage, as shown in Listing 7-2.

❶ ```
NSURLProtectionSpace *protectionSpace = [[NSURLProtectionSpace alloc] initWithHost:
    @"myhost.com" port:443 protocol:@"https" realm:nil authenticationMethod:nil];
```

❷ ```
NSURLCredential *credential = [NSURLCredential credentialWithUser:user password:
    pass persistence:NSURLCredentialPersistencePermanent];
```

❸ ```
[[NSURLCredentialStorage sharedCredentialStorage] setDefaultCredential:credential
    forProtectionSpace:protectionSpace];
```

*Listing 7-2: Setting the default credentials of a protection space*

This simply creates a protection space ❶, which includes the host, the port, the protocol, and optionally the HTTP authentication realm (if using HTTP basic authentication) and the authentication method (for example, using NTLM or another mechanism). At ❷, the example creates a credential with the username and password that it most likely received from user input. It then sets that to the default credential for this protection space at ❸, and the credential should be automatically stored in the Keychain. In the future, the app this code belongs to can read credentials with the same API, using the defaultCredentialForProtectionSpace method, as shown in Listing 7-3.

```
credentialStorage = [[NSURLCredentialStorage sharedCredentialStorage]
    defaultCredentialForProtectionSpace:protectionSpace];
```

Listing 7-3: Using the default credential for a protection space

Note, however, that credentials stored in sharedCredentialStorage are marked with the Keychain attribute kSecAttrAccessibleWhenUnlocked. If you need stricter protections, you'll need to manage Keychain storage yourself. I talk more about managing the Keychain in Chapter 13.

Also, be sure to pay attention to how you specify the value of the persistence argument when creating the credential. If you're storing in the Keychain using NSURLCredentialStorage, you can use either the NSURL-CredentialPersistencePermanent or NSURLCredentialPersistenceSynchronizable types when creating your credentials. If you're using the authentication for something more transient, the NSURLCredentialPersistenceNone or NSURL-CredentialPersistenceForSession types are more appropriate. You can find details on what each of these persistence types mean in Table 7-1.
```

Table 7-1: Credential Persistence Types

Persistence type	Meaning
NSURLCredentialPersistenceNone	Don't store the credential at all. Use this only when you need to make a single request to a protected resource.
NSURLCredentialPersistenceForSession	Persist the credential for the lifetime of your application.
NSURLCredentialPersistencePermanent	Store the credential in the Keychain.
NSURLCredentialPersistenceForSession	Persist the credential for the lifetime of your application. Use this is if you need a credential just for the time your app remains running.
NSURLCredentialPersistencePermanent	Store the credential in the Keychain. Use this when you'll want this credential on a consistent basis as long as the user has the app installed.
NSURLCredentialPersistenceSynchronizable	Store the credential in the Keychain, and allow it to be synchronized to other devices and iCloud. Use this when you want to have people transfer the credential between devices and don't have concerns about sending the credential to a third party like iCloud.

Implementing TLS Mutual Authentication with NSURLConnection

One of the best methods of performing client authentication is to use a client certificate and private key; however, this is somewhat convoluted on iOS. The basic concept is relatively simple: implement a delegate for willSendRequestForAuthenticationChallenge (formerly didReceiveAuthentication-Challenge), check whether the authentication method is NSURLAuthentication-MethodClientCertificate, retrieve and load a certificate and private key, build a credential, and use the credential for the challenge. Unfortunately, there aren't built-in Cocoa APIs for managing certificates, so you'll need to muck about with Core Foundation a fair bit, like in this basic framework:

```
- (void)connection:(NSURLConnection *) willSendRequestForAuthenticationChallenge:(
    NSURLAuthenticationChallenge *)challenge {
    if ([[[challenge protectionSpace] authenticationMethod] isEqualToString:
    NSURLAuthenticationMethodClientCertificate]) {

        SecIdentityRef identity;
        SecTrustRef trust;
❶       extractIdentityAndTrust(somep12Data, &identity, &trust);
```

```
        SecCertificateRef certificate;
❷      SecIdentityCopyCertificate(identity, &certificate);
❸      const void *certificates[] = { certificate };
❹      CFArrayRef arrayOfCerts = CFArrayCreate(kCFAllocatorDefault, certificates,
    1, NULL);

❺      NSURLCredential *cred = [NSURLCredential credentialWithIdentity:identity
        certificates:(__bridge NSArray*)arrayOfCerts
        persistence:NSURLCredentialPersistenceNone];
❻      [[challenge sender] useCredential:cred
                forAuthenticationChallenge:challenge];
    }
}
```

This example creates a `SecIdentityRef` and `SecTrustRef` so that it has destinations to pass to the `extractIdentityAndTrust` function at ❶. This function will extract the identity and trust information from a blob of PKCS #12 data (file extension *.p12*). These archive files just store a bunch of cryptography objects in one place.

The code then makes a `SecCertificateRef` into which it extracts the certificate from the identity ❷. Next, it builds an array containing the one certificate at ❸ and creates a `CFArrayRef` to hold that certificate at ❹. Finally, the code creates an `NSURLCredential`, passing in its identity and its array of certificates with only one element ❺, and presents this credential as the answer to its challenge ❻.

You'll notice some handwaving around ❶. This is because obtaining the actual certificate p12 data can happen a few different ways. You can perform a one-time bootstrap and fetch a newly generated certificate over a secure channel, generate a certificate locally, read one from the filesystem, or fetch one from the Keychain. One way to get the certificate information used in `somep12Data` is by retrieving it from the filesystem, like this:

```
NSData *myP12Certificate = [NSData dataWithContentsOfFile:path];
CFDataRef somep12Data = (__bridge CFDataRef)myP12Certificate;
```

The best place to store certificates of course is the Keychain; I'll cover that further in Chapter 13.

Modifying Redirect Behavior

By default, `NSURLConnection` will follow HTTP redirects when it encounters them. However, its behavior when this happens is, well, unusual. When the redirect is encountered, `NSURLConnection` will send a request, containing the HTTP headers as they were used in the original `NSURLHttpRequest`, to the new location. Unfortunately, this also means that the current value of your cookies for the original domain is passed to the new location. As a result, if an attacker can get your application to visit a page on your site that accepts an arbitrary URL as a place to redirect to, that attacker can steal your users' cookies, as well as any other sensitive data that your application might store in its HTTP headers. This type of flaw is called an *open redirect*.

You can modify this behavior by implementing connect:willSendRequest: redirectResponse[3] on your NSURLConnectionDelegate in iOS 4.3 and older, or on your NSURLConnectionDataDelegate in iOS 5.0 and newer.[4]

```
- (NSURLRequest *)connection:(NSURLConnection *)connection
            willSendRequest:(NSURLRequest *)request
            redirectResponse:(NSURLResponse *)redirectResponse
{
    NSURLRequest *newRequest = request;
❶   if (![[[redirectResponse URL] host] isEqual:@"myhost.com"]) {
        return newRequest;
    }

    else {
❷       newRequest = nil;
        return newRequest;
    }
}
```

At ❶, this code checks whether the domain you're redirecting to is different from the name of your site. If it's the same, it carries on as normal. If it's different, it modifies the request to be nil ❷.

TLS Certificate Pinning

In the past several years, there have been a number of troubling developments regarding certificate authorities (CAs), the entities that vouch for the TLS certificates that we encounter on a daily basis. Aside from the massive number of signing authorities trusted by your average client application, CAs have had several prominent security breaches where signing keys were compromised or where overly permissive certificates were issued. These breaches allow anyone in possession of the signing key to impersonate any TLS server, meaning they can successfully and transparently read or modify requests to the server and their responses.

To help mitigate these attacks, client applications of many types have implemented *certificate pinning*. This term can refer to a number of different techniques, but the core idea is to programmatically restrict the number of certificates that your application will trust. You could limit trust to a single

---

3. *https://developer.apple.com/library/ios/#documentation/cocoa/conceptual/URLLoadingSystem/ Articles/RequestChanges.html*

4. *https://developer.apple.com/library/ios/#documentation/Foundation/Reference/ NSURLConnectionDataDelegate_protocol/Reference/Reference.html#//apple_ref/occ/intfm/ NSURLConnectionDataDelegate/connection:willSendRequest:redirectResponse:*

CA (that is, the one that your company uses to sign its server certificates), to an internal root CA that you use to create your own certificates (the top of the chain of trust), or simply to a leaf certificate (a single specific certificate at the bottom of the chain of trust).

As part of the SSL Conservatory project, my colleague Alban Diquet has developed some convenient wrappers that allow you to implement certificate pinning in your application. (Learn more at *https://github.com/iSECPartners/ssl-conservatory*.) You could write your own wrapper or use an existing one; either way, a good wrapper can make pinning rather simple. For example, here's a look at how easy it would be to implement certificate pinning with Alban's wrapper:

```
❶ - (NSData*)loadCertificateFromFile:(NSString*)fileName {
       NSString *certPath = [[NSString alloc] initWithFormat:@"%@/%@", [[NSBundle
          mainBundle] bundlePath], fileName];
       NSData *certData = [[NSData alloc] initWithContentsOfFile:certPath];
       return certData;
   }

   - (void)pinThings {
   NSMutableDictionary *domainsToPin = [[NSMutableDictionary alloc] init];

❷ NSData *myCertData = [self loadCertificateFromFile:@"myCerts.der"];
   if (myCertData == nil) {
       NSLog(@"Failed to load the certificates");
       return;
       }

❸ [domainsToPin setObject:myCertData forKey:@"myhost.com"];

❹ if ([SSLCertificatePinning loadSSLPinsFromDERCertificates:domainsToPin] != YES) {
       NSLog(@"Failed to pin the certificates");
       return;
       }
   }
```

At ❶, this code simply defines a method to load a certificate from a DER-formatted file into an NSData object and calls this method at ❷. If this is successful, the code puts myCertData into an NSMutableDictionary ❸ and calls the loadSSLPinsFromDERCertificates method of the main SSLCertificatePinning class ❹. With these pins loaded, an app would also need to implement an NSURLConnection delegate, as shown in Listing 7-4.

```
- (void)connection:(NSURLConnection *)connection
    willSendRequestForAuthenticationChallenge:(NSURLAuthenticationChallenge *)
    challenge {

  if([challenge.protectionSpace.authenticationMethod isEqualToString:
  NSURLAuthenticationMethodServerTrust]) {

      SecTrustRef serverTrust = [[challenge protectionSpace] serverTrust];
      NSString *domain = [[challenge protectionSpace] host];
      SecTrustResultType trustResult;

      SecTrustEvaluate(serverTrust, &trustResult);
      if (trustResult == kSecTrustResultUnspecified) {

          // Look for a pinned public key in the server's certificate chain
          if ([SSLCertificatePinning verifyPinnedCertificateForTrust:serverTrust
    andDomain:domain]) {

              // Found the certificate; continue connecting
              [challenge.sender useCredential:[NSURLCredential credentialForTrust
    :challenge.protectionSpace.serverTrust] forAuthenticationChallenge:challenge];
          }
          else {
              // Certificate not found; cancel the connection
              [[challenge sender] cancelAuthenticationChallenge: challenge];
          }
      }
      else {
          // Certificate chain validation failed; cancel the connection
          [[challenge sender] cancelAuthenticationChallenge: challenge];
      }
  }
}
```

Listing 7-4: An NSURLConnection *delegate to handle certificate pinning logic*

   This simply evaluates the certificate chain presented by a remote server
and compares it to the pinned certificates included with your application.
If a pinned certificate is found, the connection continues; if it isn't, the
authentication challenge process is canceled.
   With your delegate implemented as shown, all your uses of NSURL-
Connection should check to ensure that they are pinned to a domain
and certificate pair in your predefined list. If you're curious, you can

find the rest of the code to implement your own certificate pinning at *https://github.com/iSECPartners/ssl-conservatory/tree/master/ios*. There's a fair bit of other logic involved, so I can't show all the code here.

NOTE   *If you're in a hurry, a delegate that you can just subclass is included in the SSL Conservatory sample code.*

Up to now, I've shown network security issues and solutions that revolve around NSURLConnection. But as of iOS 7, NSURLSession is preferred over the traditional NSURLConnection class. Let's take a closer look at this API.

Using NSURLSession

The NSURLSession class is generally favored by developers because it focuses on the use of network *sessions*, as opposed to NSURLConnection's focus on individual requests. While broadening the scope of NSURLConnection somewhat, NSURLSession also gives additional flexibility by allowing configurations to be set on individual sessions rather than globally throughout the application. Once sessions are instantiated, they are handed individual tasks to perform, using the NSURLSessionDataTask, NSURLSessionUploadTask, and NSURLSessionDownloadTask classes.

In this section, you'll explore some ways to use NSURLSession, some potential security pitfalls, and some security mechanisms not provided by the older NSURLConnection.

NSURLSession Configuration

The NSURLSessionConfiguration class encapsulates options passed to NSURLSession objects so that you can have separate configurations for separate types of requests. For example, you can apply different caching and cookie policies to requests fetching data of varying sensitivity levels, rather than having these policies be app-wide. To use the system policies for NSURLSession configuration, you can use the default policy of [NSURLSessionConfigurationdefaultConfiguration], or you can simply neglect to specify a configuration policy and instantiate your request object with [NSURLSessionsharedSession].

For security-sensitive requests that should leave no remnants on local storage, the configuration method ephemeralSessionConfiguration should be used instead. A third method, backgroundSessionConfiguration, is available specifically for long-running upload or download tasks. This type of session will be handed off to a system service to manage completion, even if your application is killed or crashes.

Also, for the first time, you can specify that a connection use only TLS version 1.2, which helps defend against attacks such as BEAST[5] and

---

5. *https://bug665814.bugzilla.mozilla.org/attachment.cgi?id=540839*

CRIME,[6] both of which can allow network attackers to read or tamper with your TLS connections.

NOTE *Session configurations are read-only after an* NSURLSession *is instantiated; policies and configurations cannot be changed mid-session, and you cannot swap out for a separate configuration.*

Performing NSURLSession Tasks

Let's walk through the typical flow of creating an NSURLSessionConfiguration and assigning it a simple task, as shown in Listing 7-5.

❶ NSURLSessionConfiguration *configuration = [NSURLSessionConfiguration
        ephemeralSessionConfiguration];

❷ [configuration setTLSMinimumSupportedProtocol = kTLSProtocol12];

❸ NSURL *url = [NSURL URLWithString:@"https://www.mycorp.com"];

   NSURLRequest *request = [NSURLRequest requestWithURL:url];

❹ NSURLSession *session = [NSURLSession sessionWithConfiguration:configuration
                                                delegate:self
                                                delegateQueue:nil];

❺ NSURLSessionDataTask *task = [session dataTaskWithRequest:request
                                        completionHandler:
        ^(NSData *data, NSURLResponse *response, NSError *error) {
❻   // Your completion handler block
        }];

❼ [task resume];

Listing 7-5: Creating an ephemeral NSURLConfiguration *requiring TLSv1.2*

The NSURLSessionConfiguration object is instantiated at ❶, with the specification that the connection should be ephemeral. This should prevent cached data from being written to local storage. Then, at ❷, the configuration also requires TLS version 1.2 since the developer controls the endpoint and knows that it supports that version. Next, just as with NSURLConnection, an NSURL object and an NSURLRequest object with that URL ❸ are created. With the configuration and request created, the app can then instantiate the session ❹ and assign a task to that session ❺.

---

6. *https://docs.google.com/presentation/d/11eBmGiHbYcHR9gL5nDyZChu_-lCa2GizeuOfaLU2HOU/edit?pli=1#slide=id.g1d134dff_1_222*

NSURLSessionDataTask and its siblings take a completion handler block as an argument ❻. This block asynchronously handles the server response and data you receive as a result of the task. Alternatively (or in addition), you can specify a custom delegate conforming to the NSURLSessionTaskDelegate protocol. One reason you may want to use both a completionHandler and a delegate is to have the completion handler take care of the results of the request, while the delegate manages authentication and caching decisions on a session basis instead of a task basis (I'll talk about this in the next section).

Finally, at ❼, this code sets the task running with a call to its resume method because all tasks are suspended upon creation.

Spotting NSURLSession TLS Bypasses

NSURLSession has a way to avoid TLS checks as well. Apps can just use the didReceiveChallenge delegate and pass the proposedCredential of the challenge received back as a credential for the session, as in Listing 7-6.

```
- (void)URLSession:(NSURLSession *)session didReceiveChallenge:(
      NSURLAuthenticationChallenge *)challenge completionHandler:(void (^)(
      NSURLSessionAuthChallengeDisposition disposition, NSURLCredential * credential
      ))completionHandler {

      completionHandler(NSURLSessionAuthChallengeUseCredential,
        [challenge proposedCredential]);
}
```
❶

Listing 7-6: Bypassing server verification with NSURLSession

This is another bypass that can be tricky to spot. Look for code like that at ❶, where there's a completionHandler followed by proposedCredential.

Basic Authentication with NSURLSession

HTTP authentication with NSURLSession is handled by the session and is passed to the didReceiveChallenge delegate, as shown in Listing 7-7.

```
❶ - (void)URLSession:(NSURLSession *)session didReceiveChallenge:(
      NSURLAuthenticationChallenge *)challenge completionHandler:(void (^)(
      NSURLSessionAuthChallengeDisposition, NSURLCredential *))completionHandler {
      NSString *user = @"user";
      NSString *pass = @"pass";

      NSURLProtectionSpace *space = [challenge protectionSpace];
```

```
    if ([space receivesCredentialSecurely] == YES &&
        [[space host] isEqualToString:@"myhost.com"] &&
        [[space authenticationMethod] isEqualToString:
    NSURLAuthenticationMethodHTTPBasic]) {

❷   NSURLCredential *credential =
        [NSURLCredential credentialWithUser:user
                            password:pass
                    persistence:NSURLCredentialPersistenceForSession];

❸   completionHandler(NSURLSessionAuthChallengeUseCredential, credential);
    }
}
```

Listing 7-7: A sample didReceiveChallenge *delegate*

This approach defines a delegate and a completion handler at ❶, creates an NSURLCredential at ❷, and passes that credential to the completion handler at ❸. Note that for either the NSURLConnection or NSURLSession approach, some developers forget to ensure that they're talking to the correct host or sending credentials securely. This would result in credentials getting sent to *every* URL your app loads, instead of just yours; Listing 7-8 shows an example of what that mistake might look like.

```
- (void)URLSession:(NSURLSession *)session didReceiveChallenge:(
    NSURLAuthenticationChallenge *)challenge completionHandler:(void (^)(
    NSURLSessionAuthChallengeDisposition, NSURLCredential *))completionHandler {

    NSURLCredential *credential =
        [NSURLCredential credentialWithUser:user
                            password:pass
                    persistence:NSURLCredentialPersistenceForSession];

    completionHandler(NSURLSessionAuthChallengeUseCredential, credential);
}
```

Listing 7-8: The wrong way to do HTTP auth

If you want to use persistent credentials for a dedicated endpoint, you can store them in sharedCredentialStorage as you did with NSURLConnection. When constructing your session, you can provide these credentials beforehand without having to worry about a delegate method, as shown in Listing 7-9.

```
NSURLSessionConfiguration *config = [NSURLSessionConfiguration
    defaultSessionConfiguration];
```

```
[config setURLCredentialStorage:
    [NSURLCredentialStorage sharedCredentialStorage]];

NSURLSession *session = [NSURLSession sessionWithConfiguration:config
                                               delegate:nil
                                          delegateQueue:nil];
```

Listing 7-9: Using an NSRULSessionConfiguration *to reference stored credentials*

This just creates an NSURLSessionConfiguration and specifies that it should use the shared credential storage. When you connect to a resource that has credentials stored in the Keychain, those will be used by the session.

Managing Stored URL Credentials

You've seen how to store and read credentials using sharedCredentialStorage, but the NSURLCredentialStorage API also lets you remove credentials using the removeCredential:forProtectionSpace method. For example, you may want to do this when a user explicitly decides to log out of an application or remove an account. Listing 7-10 shows a typical use case.

```
NSURLProtectionSpace *space = [[NSURLProtectionSpace alloc]
    initWithHost:@"myhost.com"
          port:443
      protocol:@"https"
         realm:nil authenticationMethod:nil];

NSURLCredential *credential = [credentialStorage
    defaultCredentialForProtectionSpace:space];

[[NSURLCredentialStorage sharedCredentialStorage] removeCredential:credential
                                       forProtectionSpace:space];
```

Listing 7-10: Removing default credentials

This will delete the credentials from your local Keychain. However, if a credential has a persistence of NSURLCredentialPersistenceSynchronizable, the credential may have been synchronized to other devices via iCloud. To remove the credentials from all devices, use the NSURLCredentialStorageRemove-SynchronizableCredentials option, as shown in Listing 7-11.

```
NSDictionary *options = [NSDictionary dictionaryWithObjects forKeys:
    NSURLCredentialStorageRemoveSynchronizableCredentials, YES];
```

```
[[NSURLCredentialStorage sharedCredentialStorage] removeCredential:credential
                                         forProtectionSpace:space
                                                  options:options];
```

Listing 7-11: Removing credentials from the local Keychain and from iCloud

At this point, you should have an understanding of the NSURLConnection and NSURLSession APIs and their basic usage. There are other network frameworks that you may encounter, which have their own behaviors and require slightly different security configuration. I'll cover a few of these now.

Risks of Third-Party Networking APIs

There are a few popular third-party networking APIs used in iOS applications, largely for simplifying various networking tasks such as multipart uploads and certificate pinning. The most commonly used one is AFNetworking,[7] followed by the now-obsolete ASIHTTPRequest.[8] In this section, I'll introduce you to both.

Bad and Good Uses of AFNetworking

AFNetworking is a popular library built on top of NSOperation and NSHTTP-Request. It provides several convenience methods to interact with different types of web APIs and perform common HTTP networking tasks.

As with other networking frameworks, one crucial task is to ensure that TLS safety mechanisms have not been disabled. In AFNetworking, TLS certificate validation can be disabled in a few ways. One is via the _AFNETWORKING_ALLOW_INVALID_SSL_CERTIFICATES flag, typically set in the *Prefix.pch* file. Another way is to set a property of AFHTTPClient, as in Listing 7-12.

```
NSURL *baseURL = [NSURL URLWithString:@"https://myhost.com"];
AFHTTPClient* client = [AFHTTPClient clientWithBaseURL:baseURL];
[client setAllowsInvalidSSLCertificate:YES];
```

Listing 7-12: Disabling TLS validation with setAllowsInvalidSSLCertificate

The last way you might see TLS validation being disabled is by changing the security policy of AFHTTPRequestOperationManager with setAllowsInvalidSSL-Certificate, as shown in Listing 7-13.

```
AFHTTPRequestOperationManager *manager = [AFHTTPRequestOperationManager manager];
[manager [securityPolicy setAllowInvalidCertificates:YES]];
```

Listing 7-13: Disabling TLS validation using securityPolicy

---

7. *https://github.com/AFNetworking/AFNetworking*

8. *https://github.com/pokeb/asi-http-request*

You'll also want to verify that the code you're examining doesn't use the `AFHTTPRequestOperationLogger` class in production versions. This logger uses `NSLog` on the backend to write requested URLs to the Apple System Log, allowing them to be seen by other applications on some iOS versions.

One particularly useful feature that AFNetworking provides is the ability to easily perform certificate pinning. You can just set the `_AFNETWORKING_PIN _SSL_CERTIFICATES_` #define in your project's .*pch* file, and set the pinning mode (`defaultSSLPinningMode`) property of your `AFHTTPClient` instance appropriately; the available modes are described in Table 7-2. You then put the certificates that you want to pin to in the bundle root, as files with a .*cer* extension.

Table 7-2: AFNetworking SSL Pinning Modes

Mode	Meaning
AFSSLPinningModeNone	Perform no certificate pinning, even if pinning is enabled. Use for debug mode if necessary.
AFSSLPinningModePublicKey	Pin to the certificate's public key.
AFSSLPinningModeCertificate	Pin to the exact certificate (or certificates) supplied. This will require an application update if a certificate is reissued.

As shown in sample code included with AFNetworking, you can examine URLs to determine whether they should be pinned. Just evaluate the scheme and domain name to see whether those domains belong to you. Listing 7-14 shows an example.

```
if ([[url scheme] isEqualToString:@"https"] &&
    [[url host] isEqualToString:@"yourpinneddomain.com"]) {
        [self setDefaultSSLPinningMode:AFSSLPinningModePublicKey];
    }

    else {
        [self setDefaultSSLPinningMode:AFSSLPinningModeNone];
    }

    return self;
}
```

Listing 7-14: Determining whether a URL should be pinned

The `else` statement is not strictly necessary because not pinning is the default, but it does provide some clarity.

Keep in mind that AFNetworking pins to all certificates provided in the bundle, but it doesn't check that the certificate common name and the hostname of the network endpoint match. This is mostly an issue if your application pins to multiple sites with different security standards. In other words, if your application pins to both *https://funnyimages.com* and *https://www.bank.com*, an attacker in possession of the *funnyimages.com* private key would be able to intercept communications from your application to *bank.com*.

Now that you've had a glimpse at how you can use and abuse the AFNetworking library, let's move on to ASIHTTPRequest.

Unsafe Uses of ASIHTTPRequest

ASIHTTPRequest is a deprecated library similar to AFNetworking, but it's a bit less complete and is based on the CFNetwork API. It should not be used for new projects, but you may find it in existing codebases where migration has been considered too expensive. When examining these codebases, the standard SSL validation bypass to look for is setValidatesSecureCertificate:NO.

You'll also want to examine *ASIHTTPRequestConfig.h* in your project to ensure that overly verbose logging is not enabled (see Listing 7-15).

```
// If defined, will use the specified function for debug logging
// Otherwise use NSLog
#ifndef ASI_DEBUG_LOG
    #define ASI_DEBUG_LOG NSLog
#endif


// When set to 1, ASIHTTPRequests will print information about what a request is
        doing
#ifndef DEBUG_REQUEST_STATUS
    #define DEBUG_REQUEST_STATUS 0
#endif


// When set to 1, ASIFormDataRequests will print information about the request body
        to the console
#ifndef DEBUG_FORM_DATA_REQUEST
    #define DEBUG_FORM_DATA_REQUEST 0
#endif


// When set to 1, ASIHTTPRequests will print information about bandwidth throttling
        to the console
#ifndef DEBUG_THROTTLING
    #define DEBUG_THROTTLING 0
#endif


// When set to 1, ASIHTTPRequests will print information about persistent
        connections to the console
#ifndef DEBUG_PERSISTENT_CONNECTIONS
    #define DEBUG_PERSISTENT_CONNECTIONS 0
#endif
```

```
// When set to 1, ASIHTTPRequests will print information about HTTP authentication
    (Basic, Digest or NTLM) to the console
#ifndef DEBUG_HTTP_AUTHENTICATION
    #define DEBUG_HTTP_AUTHENTICATION 0
#endif
```

Listing 7-15: Logging defines in ASIHTTPRequestConfig.h

If you do want to use these logging facilities, you may want to wrap them in #ifdef DEBUG conditionals, like this:

```
#ifndef DEBUG_HTTP_AUTHENTICATION
    #ifdef DEBUG
        #define DEBUG_HTTP_AUTHENTICATION 1
    #else
        #define DEBUG_HTTP_AUTHENTICATION 0
    #endif
#endif
```

This *ASIHTTPRequestConfig.h* file wraps the logging facilities inside conditionals to keep this information from leaking in production builds.

Multipeer Connectivity

iOS 7 introduced Multipeer Connectivity,[9] which allows nearby devices to communicate with each other with a minimal network configuration. Multipeer Connectivity communication can take place over Wi-Fi (either peer-to-peer or multipeer networks) or Bluetooth personal area networks (PANs). Bonjour is the default mechanism for browsing and advertising available services.

Developers can use Multipeer Connectivity to perform peer-to-peer file transfers or stream content between devices. As with any type of peer communication, the validation of incoming data from untrusted peers is crucial; however, there are also transport security mechanisms in place to ensure that the data is safe from eavesdropping.

Multipeer Connectivity sessions are created with either the initWithPeer or initWithPeer:securityIdentity:encryptionPreference: class method of the MCSession class. The latter method allows you to require encryption, as well as include a certificate chain to verify your device.

When specifying a value for encryptionPreference, your options are MCEncryptionNone, MCEncryptionRequired, and MCEncryptionOptional. Note that these are interchangeable with values of 0, 1, or 2, respectively. So while

9. *https://developer.apple.com/library/prerelease/ios/documentation/MultipeerConnectivity/Reference/ MultipeerConnectivityFramework/index.html*

values of 0 and 1 behave how you would expect if this value were a Boolean, a value of 2 is functionally equivalent to not having encryption at all.

It's a good idea to require encryption unconditionally because MCEncryptionOptional is subject to downgrade attacks. (You can find more detail in Alban Diquet's Black Hat talk on reversing the Multipeer Connectivity protocol.[10]) Listing 7-16 shows a typical invocation, creating a session and requiring encryption.

```
MCPeerID *peerID = [[MCPeerID alloc] initWithDisplayName:@"my device"];

MCSession *session = [[MCSession alloc] initWithPeer:peerID
                                 securityIdentity:nil
                          encryptionPreference:MCEncryptionRequired];
```

Listing 7-16: Creating an MCSession

When connecting to a remote device, the delegate method session:didReceiveCertificate:fromPeer:certificateHandler: is called, passing in the peer's certificate and allowing you to specify a handler method to take specific action based on whether the certificate was verified successfully.

NOTE   *If you fail to create the didReceiveCertificate delegate method or don't implement a certificateHandler in this delegate method, no verification of the remote endpoint will occur, making the connection susceptible to interception by a third party.*

When examining codebases using the Multipeer Connectivity API, ensure that all instantiations of MCSession provide an identity and require transport encryption. Sessions with any type of sensitive information should never be instantiated simply with initWithPeer. Also ensure that the delegate method for didReceiveCertificate exists and is implemented correctly and that the certificateHandler behaves properly when a peer fails certificate validation. You specifically *don't* want to see something like this:

```
- (void) session:(MCSession *)session didReceiveCertificate:(NSArray *)certificate
    fromPeer:(MCPeerID *)peerID
    certificateHandler:(void (^)(BOOL accept))certificateHandler
{
    certificateHandler(YES);
}
```

This code blindly passes a YES boolean to the handler, which you should never, ever do.

It's up to you to decide how you'd like to implement validation. Systems for validation tend to be somewhat customized, but you have a couple of basic options. You can have clients generate certificates themselves and then *trust on first use (TOFU)*, which just verifies that the certificate being

---

10. *https://nabla-c0d3.github.io/blog/2014/08/20/multipeer-connectivity-follow-up/*

presented is the same as the one shown the first time you paired with a peer. You can also implement a server that will return the public certificates of users when queried to centralize the management of identities. Choose a solution that makes sense for your business model and threat model.

Lower-Level Networking with NSStream

NSStream is suitable for making non-HTTP network connections, but it can also be used for HTTP communications with fairly little effort. For some unfathomable reason, in the transition between OS X Cocoa and iOS Cocoa Touch, Apple removed the method that allows an NSStream to establish a network connection to a remote host, getStreamsToHost. So if you want to sit around streaming things to yourself, then awesome. Otherwise, in Technical Q&A QA1652,[11] Apple describes a category that you can use to define a roughly equivalent getStreamsToHostNamed method of NSStream.

The alternative is to use the lower-level Core Foundation CFStreamCreatePairWithSocketToHost function and cast the input and output CFStreams to NSStreams, as shown in Listing 7-17.

```
NSInputStream *inStream;
NSOutputStream *outStream;

CFReadStreamRef readStream;
CFWriteStreamRef writeStream;
CFStreamCreatePairWithSocketToHost(NULL, (CFStringRef)@"myhost.com", 80, &
    readStream, &writeStream);
inStream = (__bridge NSInputStream *)readStream;
outStream = (__bridge NSOutputStream *)writeStream;
```

Listing 7-17: Casting CFStreams to NSStreams

NSStreams allow users only minor control of the characteristics of the connection, such as TCP port and TLS settings (see Listing 7-18).

```
NSHost *myhost = [NSHost hostWithName:[@"www.conglomco.com"]];

[NSStream getStreamsToHostNamed:myhost
                      port:443
              inputStream:&MyInputStream
             outputStream:&MyOutputStream];

❶ [MyInputStream setProperty:NSStreamSocketSecurityLevelTLSv1
              forKey:NSStreamSocketSecurityLevelKey];
```

Listing 7-18: Opening a basic TLS connection with NSStream

---

11. *https://developer.apple.com/library/ios/#qa/qa2009/qa1652.html*

This is the typical use of an `NSStream`: setting a host, port, and input and output streams. Since you don't have a ton of control over TLS settings, the only setting that might be screwed up is ❶, the `NSStreamSocketSecurityLevel`. You should set it to `NSStreamSocketSecurityLevelTLSv1` to ensure that you don't end up using an older, broken SSL/TLS protocol.

Even Lower-level Networking with CFStream

With `CFStreams`, the developer is given an unfortunate amount of control in TLS session negotiation.[12] See Table 7-3 for a number of `CFStream` properties that you should look for. These controls allow developers to override or disable verification of the peer's canonical name (CN), ignore expiration dates, allow untrusted root certificates, and totally neglect to verify the certificate chain at all.

Table 7-3: Horrible `CFStream` TLS Security Constants

Constant	Meaning	Default
kCFStreamSSLLevel	The protocol to be used for encrypting the connection.	negotiated[a]
kCFStreamSSLAllowsExpiredCertificates	Accept expired TLS certificates.	false
kCFStreamSSLAllowsExpiredRoots	Accept certificates that have expired root certificates in their certificate chain.	false
kCFStreamSSLAllowsAnyRoot	Whether a root certificate can be used as a TLS endpoint's certificate (in other words, a self-signed or unsigned certificate).	false
kCFStreamSSLValidatesCertificateChain	Whether the certificate chain is validated.	true
kCFStreamSSLPeerName	Overrides the hostname compared to that of the certificate's CN. If set to kCFNull, no validation is performed.	hostname
kCFStreamSSLIsServer	Whether this stream will act as a server.	false
kCFStreamSSLCertificates	An array of certificates that will be used if kCFStreamSSLIsServer is true.	none

a. The default constant is `kCFStreamSocketSecurityLevelNegotiatedSSL`, which negotiates the strongest method available from the server.

12. *https://developer.apple.com/library/mac/#documentation/CoreFoundation/Reference/CFSocketStreamRef/Reference/reference.html*

You probably shouldn't be using these security constants at all, but if you must use TLS CFStreams, just do it the right way. It's simple! Provided that you're not creating a network server within the app itself (which is a pretty rare usage of CFStream in an iOS app), there are two steps you should follow:

1. Set kCFStreamSSLLevel to kCFStreamSocketSecurityLevelTLSv1.

2. Don't mess with anything else.

Closing Thoughts

You've looked at quite a number of ways for apps to communicate with the outside world and the incorrect ways those things can be implemented. Let's now turn our attention to communication with other applications and some of the pitfalls that can happen when shuffling data around via IPC.

8

INTERPROCESS COMMUNICATION

Interprocess communication (IPC) on iOS is, depending on your perspective, refreshingly simple or horribly limiting. I mostly consider it to be the former. While Android has flexible IPC mechanisms such as Intents, Content Providers, and Binder, iOS has a simple system based on two components: message passing via URLs and application extensions. The message passing helps other applications and web pages invoke your application with externally supplied parameters. Application extensions are intended to extend the functionality of the base system, providing services such as sharing, storage, and the ability to alter the functionality of the Today screen or keyboard.

In this chapter, you'll learn about the various ways you can implement IPC on iOS, how people commonly get IPC wrong, and how to work around some of the limitations imposed by this system without compromising user security.

URL Schemes and the openURL Method

The official IPC mechanism available to iOS application developers is via URL schemes, which are similar to protocol handlers such as `mailto:` on a desktop system.

On iOS, developers can define a URL scheme that they want their application to respond to, and other applications (or web pages, importantly) can invoke the application by passing in arguments as URL parameters.

Defining URL Schemes

Custom URL schemes are described in a project's *Info.plist* file. To add a new scheme, you can use Xcode's plist editor, shown in Figure 8-1.

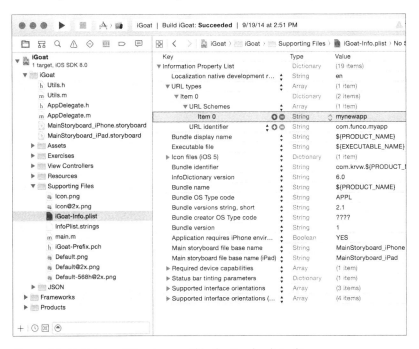

Figure 8-1: Defining a URL scheme within the Xcode plist editor

First, you add the URL types key, which will create a subkey, called Item 0. A subkey will automatically be created for the URL identifier, which should be populated with a reverse DNS notated string such as `com.mycompany.myapp`. Then, you create a new subkey of Item 0, which is the URL Schemes key. Under Item 0, which was created under URL Schemes, enter the scheme you want other applications to call your application by. For example, entering `mynewapp` here makes it so that your application will respond to *mynewapp://* URLs.

You can also define these URL schemes manually within the plist file using an external editor, as shown in Listing 8-1.

```
<?xml version="1.0" encoding="UTF-8"?>
<!DOCTYPE plist PUBLIC "-//Apple//DTD PLIST 1.0//EN" "http://www.apple.com/DTDs/
    PropertyList-1.0.dtd">
<plist version="1.0">
<dict>
  <key>CFBundleDevelopmentRegion</key>
  <string>en</string>
  <key>CFBundleURLTypes</key>
  <array>
    <dict>
      <key>CFBundleURLSchemes</key>
      <array>
        <string>com.funco.myapp</string>
      </dict>
    </array>
    <key>CFBundleDisplayName</key>
    <string>${PRODUCT_NAME}</string>
    <key>CFBundleExecutable</key>
    <string>${EXECUTABLE_NAME}</string>
```

Listing 8-1: URL schemes as shown in the plist

The bold lines indicate the additions to the original plist after the creation of the URL scheme in Figure 8-1. Learn what's in this file so that you can quickly grep for the information you need when examining a new and foreign codebase. When you're hunting for a custom URL scheme, you should look for the CFBundleURLSchemes key.

Once you've defined a URL scheme or discovered a URL scheme you want to interact with, you'll need to implement code to make or receive IPC calls. Thankfully, this is fairly simple, but there are a few pitfalls to watch out for. You'll take a look at them now.

Sending and Receiving URL/IPC Requests

To send a message via a URL scheme, you simply create an NSURL object containing an NSString representing the URL you want to call and then invoke the openURL: method [UIApplication sharedApplication]. Here's an example:

```
NSURL *myURL = [NSURL URLWithString:@"someotherapp://somestuff?someparameter=avalue
    &otherparameter=anothervalue"];

[[UIApplication sharedApplication] openURL:myURL];
```

Keys and values for the receiving application are passed in as they would be in an HTTP URL, using ? to indicate parameters and & to separate key-value pairs. The only exception is that there doesn't need to be any text before the ? because you're not talking to a remote site.

The receiving application can then extract any component of the URL with the standard NSURL object properties,[1] such as host (somestuff in my example), or the query (your key-value pairs).

Validating URLs and Authenticating the Sender

When the receiving application is invoked with its custom URL scheme, it has the option to verify that it wants to open the URL to begin with, using the application:didFinishLaunchingWithOptions: method or application:will-FinishLaunchingWithOptions: method. Applications typically use the former, as in Listing 8-2.

```
- (BOOL)application:(UIApplication *)application didFinishLaunchingWithOptions:(
    NSDictionary *)launchOptions {

    if ([launchOptions objectForKey:UIApplicationLaunchOptionsURLKey] != nil) {
        NSURL *url = (NSURL *)[launchOptions valueForKey:
    UIApplicationLaunchOptionsURLKey];
        if ([url query] != nil) {
            NSString *theQuery = [[url query]
    stringByReplacingPercentEscapesUsingEncoding:NSUTF8StringEncoding];
            if (![self isValidQuery:theQuery]) {
                return NO;
            }
            return YES;
        }
    }
}
```

Listing 8-2: Validating URLs within didFinishLaunchingWithOptions

If YES is returned, the openURL method will be called with the supplied URL. In the openURL method, the data passed (if any) is parsed and openURL makes decisions as to how the app will behave in response. The method is also where you can make decisions based on the application that called your app. Listing 8-3 shows what an openURL method might look like.

```
- (BOOL)application:(UIApplication *)application openURL:(NSURL *)url
    sourceApplication:(NSString *)sourceApplication annotation:
    (id)annotation {

❶     if ([sourceApplication isEqualToString:@"com.apple.mobilesafari"]) {
```

---

1. *https://developer.apple.com/library/mac/documentation/Cocoa/Reference/Foundation/Classes/ NSURL_Class/index.html#//apple_ref/doc/uid/20000301-SW21*

```
            NSLog(@"Loading app from Safari");
            return NO; // We don't want to be called by web pages
        }
        else {
❷          NSString *theQuery = [[url query]
        stringByReplacingPercentEscapesUsingEncoding:NSUTF8StringEncoding];
❸          NSArray *chunks = [theQuery componentsSeparatedByString:@"&"];
            for (NSString* chunk in chunks) {
❹              NSArray *keyval = [chunk componentsSeparatedByString:@"="];
❺              NSString *key = [keyval objectAtIndex:0];
                NSString *value = [keyval objectAtIndex:1];
❻              // Do something with your key and value
                --snip--
                return YES;
            }
        }
    }
```

Listing 8-3: Parsing the data received by openURL

At ❶, the method examines the source application to see whether it comes from the bundle ID that identifies Mobile Safari; since this application is meant to take input only from other applications, it returns NO. If your app is meant to be opened only by a specific application, you could restrict it to one valid bundle ID.

At ❷, the input is unescaped, in case there are URL-encoded characters in it (such as %20 for a space). At ❸ and ❹, individual key-value pairs are separated out and broken down further into key-value pairs. The first key-value pair is grabbed at ❺, and it is parsed to inform whatever logic might be written at ❻.

The parsing and validation of the actual query string will depend on what type of data you're receiving. If you're expecting a numeric value, you can also use a regular expression to ensure that the string contains only numbers. Here's an example of a check you might add to your openURL method:

```
NSCharacterSet* notNumeric = [[NSCharacterSet decimalDigitCharacterSet] invertedSet
    ];
if ([value rangeOfCharacterFromSet:notDigits].location != NSNotFound) {
    return NO; // We didn't get a numeric value
}
```

Just validate any parameters received via URL-based IPC to ensure that they contain only the type of data you expect. If you use these parameters to form database queries or change the content of the HTML, make extra sure you're sanitizing the data and integrating the content properly. I'll talk more about this in Chapter 12.

Watch for Deprecated Validation Code

Note that you may sometimes see the deprecated (yet more sensibly named) `handleOpenURL` method used in some codebases; see Listing 8-4 for an example.

```
- (BOOL)application:(UIApplication *)application handleOpenURL:(NSURL *)url
```

Listing 8-4: Deprecated method for handling received URLs

Using `handleOpenURL` is undesirable in many cases because the method blindly opens any URL given to it, and it gives you no way to identify where the URL came from. Of course, verifying the source application provides only limited guarantees.

How Safe Is Sender Validation?

Given what I've discussed in this section, you may well wonder whether you can trust the value of the `sourceApplication` parameter at all. Good question! While the sender check is merely a string comparison and is not directly cryptographic, Apple does ensure that all app IDs submitted to the App Store are unique: first come, first served. On a jailbroken device, however, you can't guarantee this uniqueness, so be wary of blindly trusting a URL just because it claims to come from a particular application.

URL Scheme Hijacking

The relatively simple system of URL scheme definition that I described has a potential problem. What if another application tries to register your URL scheme? In the case of Apple's built-in applications, other applications won't be able to successfully register a duplicate scheme. For everyone else, though, the resultant behavior is . . . undefined. Just ask Apple:

> If more than one third-party app registers to handle the same URL scheme, there is currently no process for determining which app will be given that scheme.[2]

In other words, you face two unpleasant possibilities. First, a malicious application installed before your application could register your URL scheme and retain it after your application is installed. Or, a malicious application installed after your application could successfully register your URL scheme, effectively hijacking it from your application. Either situation can result in data intended for your application going to a malicious third-party app. What can you do? I'll let you know once I figure that out.

In recent versions of iOS, however, alternative mechanisms for passing data between applications have been made available, each appropriate for

---

2. *http://developer.apple.com/library/ios/documentation/iPhone/Conceptual/ iPhoneOSProgrammingGuide/iPhoneAppProgrammingGuide.pdf* (page 99)

different circumstances. These may be a better fit for your app than openURL. Let's look at a few of these newer methods now.

Universal Links

URL scheme hijacking is one of the reasons that Apple introduced *Universal Links* in iOS 9. Universal Links are a way to effectively provide deep linking in an iOS application and integration between websites and mobile applications. For example, imagine you've published an instant messaging application called HoopChat. If a user visits a website that has a "Message me in HoopChat!" button, this could link to a URL like *https://www.hoopchat.com/im/send/?id=356372.* If the user clicks this link and has your application installed, the link would open directly in your application, where the app could create a new message to the person with the user ID of 356372. If the user doesn't have the application installed, the same URL would be viewed in Mobile Safari, which would take you to a web-based UI to send a message.

Behind the scenes, the way this works is that the application has an entitlement that specifies how it handles links to particular domains, as shown in Figure 8-2.

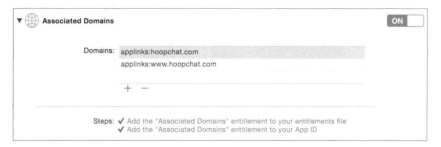

Figure 8-2: Enabling Universal Links under Associated Domains in Xcode

When one of these domains is visited in Mobile Safari, a file called *apple-app-site-association* is downloaded from the web server. This takes the form of a signed blob of JSON, as in Listing 8-5.

```
{
    "applinks": {
        "apps": [],
        "details": {
❶          "FAFBQM3A4N.com.hoopchat.messenger": {
❷              "paths": [ "*" ]
            }
        }
    }
}
```

Listing 8-5: Format of the apple-app-site-association *file*

This file specifies the developer team ID, the bundle identifier (shown at ❶), and the URL paths that should be handled by the app (as opposed to the main website). In this case, all URLs should be handled by the app if it's installed, so the file gives a value of * at ❷.

As mentioned, this blob needs to be signed; the signing key is actually the private key to your production SSL certificate. If you have the private and public keys to your website, your JSON file can be signed from the command line, as shown in Listing 8-6.

```
openssl smime \
    -sign \
    -nodetach \
❶    -in "unsigned.json" \
❷        -out "apple-app-site-association" \
    -outform DER \
❸        -inkey "private-key.pem" \
❹        -signer "certificate.pem"
```

Listing 8-6: Signing the apple-app-site-association *file*

This example uses the `openssl` utility, providing it with the unsigned JSON file at ❶ and the output filename at ❷. At ❸ and ❹, a key pair is provided. If your key is protected by a passphrase, you'd enter that when prompted, and you'd receive a valid *apple-app-site-association* file as the output. This file would then be uploaded to the web root of your website, where iOS would fetch it over HTTPS to determine whether to pass the URL to your app. Within the application, logic as to what action your app will take upon receiving a universal link will depend on what you implement in the `application:continueUserActivity:restorationHandler:` method of your application delegate.

This universal linking approach is preferable to custom URL handling schemes for a few reasons. First, Universal Links isn't subject to URL scheme hijacking; only your website, authenticated over HTTPS, can specify what URLs will be opened in your application, and those calls can't be sent to a separate bundle ID. Second, the links should work regardless of whether an app is installed. In earlier versions of iOS, you'd just get an error saying that the scheme isn't recognized. With Universal Links, if the app isn't installed, you'll be sent to the equivalent on the website. Finally, Universal Links provide some privacy protections by preventing applications from enumerating what applications are present on a device. (Apps could previously use the `canOpenURL` method to enumerate installed applications; with Universal Links, no such mechanism exists.)

Now that you've seen how you can control interactions with your own application, let's take a look at some ways to more deeply integrate your application with popular apps and services using `UIActivity`.

Sharing Data with UIActivity

In iOS 6, Apple started allowing third-party applications to share information through a set of predefined methods, such as sending data via an email or posting to Facebook. This limited form of IPC allows developers to implement the most basic sharing functionality. You can get an idea of the types of data this is useful for by examining the following UIActivity types:

- UIActivityTypePostToFacebook
- UIActivityTypePostToTwitter
- UIActivityTypePostToWeibo
- UIActivityTypePostToTencentWeibo
- UIActivityTypePostToFlickr
- UIActivityTypePostToVimeo
- UIActivityTypeMessage
- UIActivityTypeMail
- UIActivityTypePrint
- UIActivityTypeCopyToPasteboard
- UIActivityTypeAssignToContact
- UIActivityTypeSaveToCameraRoll
- UIActivityTypeAddToReadingList
- UIActivityTypeAirDrop

To share via UIActivity, just create a UIActivityViewController and pass it data such as text, a URL, an image, and so forth, as shown in Listing 8-7.

```
NSString *text = @"Check out this highly adequate iOS security resource";
NSURL *url = [NSURL URLWithString:@"http://nostarch.com/iossecurity/"];

UIActivityViewController *controller = [[UIActivityViewController alloc]
                    initWithActivityItems:@[text, url]
                    applicationActivities:nil];

[navigationController presentViewController:controller animated:YES completion:nil
    ];
```

Listing 8-7: Instantiating a UIActivityViewController

Here, a UIActivityViewController called controller is passed some text and a URL. If certain modes of sharing aren't appropriate for the data, you can exclude them. For example, if you want to ensure that users can only mail or print your content but not post to social networking sites, you can tell UIActivityViewController to exclude all other known types of sharing, as in Listing 8-8.

```
[controller setExcludedActivityTypes:@[UIActivityTypePostToFacebook,
                                        UIActivityTypePostToTwitter
                                        UIActivityTypePostToWeibo
                                        UIActivityTypePostToTencentWeibo
                                        UIActivityTypePostToFlickr
                                        UIActivityTypePostToVimeo
                                        UIActivityTypeMessage
                                        UIActivityTypeCopyToPasteboard
                                        UIActivityTypeAssignToContact
                                        UIActivityTypeSaveToCameraRoll
                                        UIActivityTypeAddToReadingList
                                        UIActivityTypeAirDrop];
```

Listing 8-8: Excluding certain types of sharing activities

This exclusion approach is, unfortunately, not convenient or thorough, and any sharing types added in future versions of iOS will be included by default. If it's important to disable parts of the sharing UI, be sure that you test with the most recent versions of iOS before they reach the general public.

In addition to URL schemes and UIActivity methods, there's one more way to handle IPC in iOS: through extensions.

Application Extensions

In iOS 8 and later, developers can write various *extensions*, which behave like specialized forms of IPC. The extensions allow you to present data to other applications, have applications share data through your app, or alter system behavior. Table 8-1 shows the various kinds of *extension points* you can code for. An extension point defines what component of the OS the extension will have access to and how it will need to be coded.

Table 8-1: Extension Points

Type	Function
Today	Manipulates widgets in the Today view of the Notification Center
Share	Allows data to be sent to your app via Share buttons
Action	Reads or manipulates data to be returned to the host app
Photo	Provides methods to manipulate photos within the Photos app
Document Provider	Allows access to a library of files
Keyboard	Provides a custom keyboard

While app extensions aren't applications, they are required to come bundled with an application, referred to as the *containing app*. Third-party applications that use an extension (called *host apps*) can communicate with the extension bundled in the containing app, but the containing app itself does not directly talk to the extension. Apple also specifically excludes some functions from being accessible via extensions, such as using the HealthKit API, receiving AirDrop data, or accessing the camera or microphone.

Extensions can be implemented in many ways, and they can be treated as applications in and of themselves. As shown in Figure 8-3, extensions are created as their own applications within Xcode.

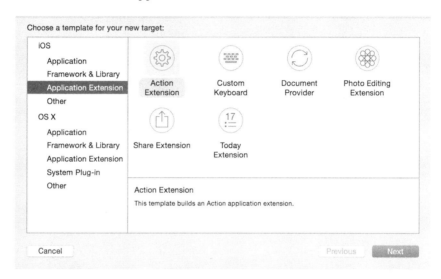

Figure 8-3: Adding a new extension target to a project

For this book, however, let's home in on the most important aspects to check from a security perspective.

Checking Whether an App Implements Extensions

First, you can easily determine whether the app you're examining implements an extension by searching for the NSExtensionPointIdentifier inside property lists. To search for that property, you can execute the following command in the project's *root* directory:

```
$ find . -name "*.plist" |xargs grep NSExtensionPointIdentifier
```

This greps all *.plist* files in the directory for NSExtensionPointIdentifier. You can also search for the property by checking the *.plist* file within Xcode, as in Figure 8-4.

Key	Type	Value
▼ Information Property List	Dictionary	(11 items)
Localization native development r...	String	en
Bundle display name	String	shareshare
Executable file	String	$(EXECUTABLE_NAME)
Bundle identifier	String	com.thiel.myApp.$(PRODUCT_NAME:rfc1034identifier)
InfoDictionary version	String	6.0
Bundle name	String	$(PRODUCT_NAME)
Bundle OS Type code	String	XPC!
Bundle versions string, short	String	1.0
Bundle creator OS Type code	String	????
Bundle version	String	1
▼ NSExtension	Dictionary	(3 items)
▼ NSExtensionAttributes	Dictionary	(1 item)
NSExtensionActivationRule	String	TRUEPREDICATE
NSExtensionMainStoryboard	String	MainInterface
NSExtensionPointIdentifier	String	com.apple.share-services

Figure 8-4: The Info.plist *of a newly created extension, viewed in Xcode*

An extension's *Info.plist* file will contain the type of extension being used, as well as optional definitions of the types of data that the extension is designed to handle. If you find the `NSExtensionPointIdentifier` property defined, you should dig in to the project and find the view controller for the defined extension.

Restricting and Validating Shareable Data

For share and action extensions, you can define an `NSExtensionActivationRule`, which contains a dictionary of data types that your application is restricted to handling (see Figure 8-5).

▼ NSExtension	Dictionary	(3 items)
▼ NSExtensionAttributes	Dictionary	(1 item)
▼ NSExtensionActivationRule	Dictionary	(1 item)
NSExtensionActivationRuleSupportsWebURLWithMaxCount	Number	1
NSExtensionMainStoryboard	String	MainInterface
NSExtensionPointIdentifier	String	com.apple.share-services

Figure 8-5: Activation rules in an extension's .plist *file, viewed in Xcode*

This dictionary will be evaluated to determine what data types your extension allows and the maximum number of these items you'll accept. But apps aren't limited to accepting predefined types of data; they can also implement custom `NSPredicate`s to define their own rules for what they'll accept. If this is the case, you'll see the `NSExtensionActivationRule` represented as a string rather than a numeric value.

If you know you're dealing with predefined data types, however, keep the following predefined activation rules in mind:

- `NSExtensionActivationSupportsAttachmentsWithMaxCount`
- `NSExtensionActivationSupportsAttachmentsWithMinCount`
- `NSExtensionActivationSupportsFileWithMaxCount`

- NSExtensionActivationSupportsImageWithMaxCount

- NSExtensionActivationSupportsMovieWithMaxCount

- NSExtensionActivationSupportsText

- NSExtensionActivationSupportsWebURLWithMaxCount

- NSExtensionActivationSupportsWebPageWithMaxCount

Because extensions can often receive unknown and arbitrary kinds of data, it's important to ensure that your extension performs correct validation in the isContentValid method of its view controller, particularly in share or action extensions. Examine the logic in your app's implementation of this method and determine whether the app is performing the necessary validation required.

Typically, an extension will examine the NSExtensionContext (which is passed in by the host app when it calls beginRequestWithExtensionContext), as in Listing 8-9.

```
NSExtensionContext *context = [self extensionContext];
NSArray *items = [context inputItems];
```

Listing 8-9: Creating an array of NSExtensionItems from the NSExtensionContext

This will give an array of NSExtensionItem objects, and each object will contain a different type of data passed in by the host app, such as images, URLs, text, and so on. Each of these items should be examined and validated before you use them to perform actions or allow the user to post the data.

Preventing Apps from Interacting with Extensions

Keyboard extensions have the unique characteristic that they read every keystroke that a user enters into them. Different third-party keyboards may have various degrees of keystroke logging to help with things such as auto-completion or sending data to a remote web service for processing. It's also possible that an actively malicious keyboard could be distributed, working as a pure keylogger. If your application accepts security-sensitive data via the keyboard, you may want to prevent the use of third-party keyboards with your application. You can do this with the shouldAllowExtensionPointIdentifier delegate method, as shown in Listing 8-10.

```
- (BOOL)application:(UIApplication *)application
    shouldAllowExtensionPointIdentifier:(NSString *)extensionPointIdentifier {
    if ([extensionPointIdentifier isEqualToString:
    UIApplicationKeyboardExtensionPointIdentifier]) {
        return NO;
    }
}
```

```
        return YES;
    }
```

Listing 8-10: The `shouldAllowExtensionPointIdentifier` *delegate method*

This code simply examines the value of `extensionPointIdentifier` and returns `NO` if it matches the constant `UIApplicationKeyboardExtensionPoint-Identifier`. Note that currently third-party keyboards are the only extensions that can be disabled in this fashion.

You've seen the best ways to implement IPC, so to close the chapter, I'll walk you through one approach to IPC that you may see in the wild that doesn't work out so well.

A Failed IPC Hack: The Pasteboard

There have been occasional reports of people abusing the `UIPasteboard` mechanism as a kind of IPC channel. For example, some try using it to transfer a user's data from a free version of an application to a "pro" version, since there's no way the newly installed application can read the old application's data. Don't do that!

An OAuth library designed to work with Twitter[3] uses the general pasteboard as a mechanism to shuffle authentication information from a web view to the main part of the app, as in this example:

```
   - (void) pasteboardChanged: (NSNotification *) note {
❶     UIPasteboard *pb = [UIPasteboard generalPasteboard];

       if ([note.userInfo objectForKey:UIPasteboardChangedTypesAddedKey] == nil)
        return;
       NSString *copied = pb.string;

       if (copied.length != 7 || !copied.oauthtwitter_isNumeric) return;
❷     [self gotPin:copied];
   }
```

After reading data from the general pasteboard at ❶, this library validates the data and sends it to the `gotPin` method at ❷.

But the general pasteboard is shared among all applications and can be read by any process on the device. This makes the pasteboard a particularly bad place to store anything even resembling private data. I'll go into more detail on the pasteboard in Chapter 10, but for now, ensure that the app you're examining isn't putting anything on the pasteboard that you wouldn't want every other app to know about.

---

3. *https://github.com/bengottlieb/Twitter-OAuth-iPhone*

Closing Thoughts

While IPC in iOS appears limited at first, there are ample opportunities for developers to fail to parse externally supplied input, create new data leaks, and even potentially send data to the wrong app. Ensure that sharing is appropriately limited, received data is validated, sending applications are verified, and unencrypted data isn't passed by simply trusting that the receiving URL handler is the one you would expect.

9

IOS-TARGETED WEB APPS

Since the introduction of third-party developer APIs for iOS, the Web has been an important component of iOS applications. Originally, the APIs were entirely web-based. While this potentially made life easier for people with no Objective-C or Cocoa experience, it severely limited what non-Apple applications could do and relegated them to a second-class status. They had no access to native capabilities of the phone, such as geolocation, and were available only within the browser instead of on the home screen.

While things have changed drastically since that time, the need to integrate with web applications from iOS has not. In this chapter, you'll take a closer look at the connections between native iOS applications and web applications: how web applications are interacted with, what native iOS APIs can be exposed to web apps, and the risks of various approaches.

Using (and Abusing) UIWebViews

Developers use web views to render and interact with web content in iOS applications because they are simple to implement and provide browser-like functionality. Most web views are instances of the UIWebView class, which uses the WebKit rendering engine[1] to display web content. Web views are often used either to abstract portions of the application so they can be shared between different mobile app platforms or simply to offload more logic to the web application, often because of more in-house expertise in web application programming than iOS. They're also frequently used as a way to view links to third-party web content without having to leave the application and spawn Safari. For example, when you click an article in a Facebook feed, the content is rendered in the Facebook app.

Starting with iOS 8, the `WKWebView` framework was introduced. This framework gives developers some additional flexibility as well as access to Apple's high-performance Nitro JavaScript engine, which increases the performance of apps that use web views significantly. Since you'll be seeing `UIWebView` for some time to come, you'll examine both APIs in this chapter, beginning with `UIWebView`.

Working with UIWebViews

Web views shift some portion of the application logic to a remote web API or application. As such, developers have less control over the behavior of web views than a fully native iOS application would allow, but there are a few controls you can put in place to bend web views to your will.

By implementing the `shouldStartLoadWithRequest` method of the protocol `UIWebViewDelegate`,[2] you can make decisions about all the URLs opened via web views before actually allowing them to be opened. For example, to limit the attack surface, you can limit all requests so that they go only to HTTPS URLs or only to particular domains. If you want to ensure that your application will never load non-HTTPS URLs, you can do something like the example shown in Listing 9-1.

```
- (BOOL)webView:(UIWebView*)webView shouldStartLoadWithRequest:(NSURLRequest*)
      request
 navigationType:(UIWebViewNavigationType)navigationType {

     NSURL *url = [request URL];

❶   if ([[url scheme] isEqualToString:@"https"]) {

        if ([url host] != nil) {
            NSString *goodHost = @"happy.fluffy.bunnies.com";

❷          if ([[url host] isEqualToString:goodHost]) {
               return YES;
           }
```

1. *https://www.webkit.org/*

2. *http://developer.apple.com/library/ios/documentation/uikit/reference/UIWebViewDelegate_Protocol/ Reference/Reference.html*

```
        }
    }
    return NO;
}
```

Listing 9-1: Rejecting non-HTTPS URLs and unknown hostnames

This example uses two different attributes of the NSURL associated with the NSURLRequest being loaded. At ❶, the scheme property of the URL is checked to see whether it matches the specified scheme, https. At ❷, the host property is compared to a single whitelisted domain: *happy.fluffy.bunnies.com.* These two restrictions limit the application's web views access to only your domain—rather than anything that might be attacker-controlled—and ensure that requests are always transmitted over HTTPS, keeping their contents safe from network attackers.

Web views may seem like the way to go because you can reuse codebases across platforms and still have some level of control over the local system. However, web views do have some serious security implications. One limitation is the inability to upgrade the WebKit binary shipped with UIWebView. WebKit is prepackaged with new versions of iOS and does not get updated out-of-band from the main OS. This means that any WebKit vulnerabilities that are discovered remain exploitable until a new version of iOS is released.

Another major part of using web views securely involves handling your cached data properly, which I'll discuss in the next section.

Executing JavaScript in UIWebViews

The web view JavaScript engine is known as JavaScriptCore, also marketed as Nitro by Apple. While the new WKWebView class improves JavaScript support (see "Enter WKWebView" on page 158), the implementation of JavaScript-Core as used in UIWebView has a couple of shortcomings when compared with JavaScript engines in modern browsers. The main limitation is the lack of just-in-time (JIT) compilation.

UIWebView JavaScript execution also limits total allocations to 10MB and runtime to 10 seconds, at which point execution will be immediately and unequivocally halted. Despite these shortcomings, applications can execute a limited amount of JavaScript by passing the script to stringByEvaluating-JavaScriptFromString, as illustrated in Listing 9-2.

```
[webView stringByEvaluatingJavaScriptFromString:@"var elem =
    document.createElement('script');"
        "elem.type = 'text/javascript';"
        "elem.text = 'aUselessFunc(name) {"
        "    alert('Ohai!'+name);"
        "};"
```

```
            "document.getElementById('head').appendChild(elem);"];
[webView stringByEvaluatingJavaScriptFromString:@"aUselessFunc('Mitch');"];
```

Listing 9-2: Injecting JavaScript into the web view

The `stringByEvaluatingJavaScriptFromString` method takes a single argument, which is a blob of JavaScript, to insert into the view. Here, the element `elem` is created, a simple function to spawn an alert box is defined, and the function is inserted into the web view. Now, the newly defined function can be called with subsequent calls to `stringByEvaluatingJavaScriptFromString`.

Do note, however, that allowing dynamic JavaScript execution within your apps exposes your users to the JavaScript injection attacks. As such, this functionality should be used judiciously, and developers should never reflect untrusted input into dynamically generated scripts.

You'll learn more about JavaScriptCore in the next section, where I discuss ways to get around the `UIWebView` shortcomings I've described so far.

Rewards and Risks of JavaScript-Cocoa Bridges

To overcome the limitations of `UIWebView`, various workarounds have been used to expose more native functionality to web-based applications. For example, the Cordova development framework uses a clever (or dangerous) web view implementation to access Cocoa APIs that allow the use of the camera, accelerometer, geolocation capabilities, address book, and more.

In this section, I'll introduce you to some popular JavaScript-Cocoa bridges, provide examples of how you'd see them used in the wild, and discuss some security risks they pose.

Interfacing Apps with JavaScriptCore

Prior to iOS 7, [`UIWebView stringByEvaluatingJavaScriptFromString:`] was the only way to invoke JavaScript from inside an application. However, iOS 7 shipped with the JavaScriptCore framework, which has full support for bridging communications between native Objective-C and a JavaScript runtime. The bridge is created via the new `JSContext` global object, which provides access to a JavaScript virtual machine for evaluating code. The Objective-C runtime can also obtain strong references to JavaScript values via `JSValue` objects.

You can use JavaScriptCore to interface with the JavaScript runtime in two fundamental ways: by using inline blocks or by directly exposing Objective-C objects with the `JSExport` protocol. I'll briefly cover how both methods work and then discuss security concerns introduced by this new attack surface.

Directly Exposing Objective-C Blocks

One use of Objective-C blocks is to provide a simple mechanism to expose Objective-C code to JavaScript. When you expose an Objective-C block to

```
```

JavaScript, the framework automatically wraps it with a callable JavaScript function, which allows you to then call the Objective-C code directly from JavaScript. Let's look at an example—albeit a contrived one—in Listing 9-3.

```
JSContext *context = [[JSContext alloc] init];
❶ context[@"shasum"] = ^(NSString *data, NSString *salt) {
    const char *cSalt  = [salt cStringUsingEncoding:NSUTF8StringEncoding];
    const char *cData = [data cStringUsingEncoding:NSUTF8StringEncoding];
    unsigned char digest[CC_SHA256_DIGEST_LENGTH];
    CCHmac(kCCHmacAlgSHA256, cSalt, strlen(cSalt), cData, strlen(cData),
     digest);
    NSMutableString *hash = [NSMutableString stringWithCapacity:
     CC_SHA256_DIGEST_LENGTH];
    for (int i = 0; i < CC_SHA256_DIGEST_LENGTH; i++) {
        [hash appendFormat:@"%02x", digest[i]];
    }
    return hash;
};
```

Listing 9-3: Exposing an Objective-C block to JavaScript

Here, a block (you can see it defined by the ^ operator at ❶) is exposed that accepts a password and a salt from JavaScript and uses the Common-Crypto framework to create a hash. This block can then be accessed directly from JavaScript to create the user's password hash, as shown in Listing 9-4.

```
var password = document.getElementById('password');
var salt = document.getElementById('salt');
var pwhash = shasum(password, salt);
```

Listing 9-4: JavaScript call to exposed Objective-C block

This technique lets you utilize the Cocoa Touch APIs and avoid re-implementing difficult and easily botched operations such as encryption or hashing.

Blocks are the simplest way to expose Objective-C code to JavaScript, but they have a few drawbacks. For instance, all the bridged objects are immutable, so changing the value of an Objective-C variable won't affect the JavaScript variable that it is mapped to. However, if you do need to share objects between both execution contexts, you can also expose custom classes using the JSExport protocol.

Connecting Objective-C and JavaScript with JSExport

The JSExport protocol allows applications to expose entire Objective-C classes and instances to JavaScript and operate on them as if they were JavaScript objects. Additionally, the references to their Objective-C counterparts are strong, meaning modifications to an object in one environment are reflected in the other. Defining variables and methods within a protocol that inherits JSExport signals to JavaScriptCore that those elements can be accessed from JavaScript, as illustrated in Listing 9-5.

```
@protocol UserExports <JSExport>

//exported variables
@property NSString *name;
@property NSString *address;

//exported functions
- (NSString *) updateUser:(NSDictionary *)info;
@end
```

Listing 9-5: Exposing variables and methods using a whitelist approach

Thanks to that JSExport protocol declaration, JavaScript has access to the variables name and address and the function updateUser. Apple has made exposing such objects to JavaScriptCore extremely easy, which means it can also be extremely easy for developers to inadvertently expose all kinds of unintended functionality. Luckily, this bridge follows an entirely opt-in model: only members you actually define in the protocol itself are exposed. Unless explicitly whitelisted in the protocol definition, any additional declarations made in the class interface are hidden, as in Listing 9-6.

```
@interface User : NSObject <UserExports> ❶

// non-exported variable
@property NSString *password;

// non-exported method declaration
- (BOOL) resetPassword;
@end
```

Listing 9-6: Elements declared outside the protocol definition are inacessible in JavaScript

The User interface inherits from UserExports at ❶, so it also inherits from JSExport. But the password property and the resetPassword method aren't declared inside UserExports, so they won't be exposed to JavaScript.

Now that JavaScriptCore knows about your UserExports protocol, it can create an appropriate wrapper object when you add an instance of it to a JSContext, as in the next listing:

```
❶  JSContext *context = [[JSContext alloc] init];
❷  User *user= [[User alloc] init];
```

```
[user setName:@"Winston Furchill"];
[user setValue:24011965];
[user setHiddenName:@"Enigma"];
❸ context[@"user"] = user;
❹ JSValue val = [context evaluateScript:@"user.value"];
❺ JSValue val = [context evaluateScript:@"user.hiddenName"];
NSLog(@"value: %d", [val toInt32]); // => 23011965
NSLog(@"hiddenName: %@", [val toString]); // => undefined
```

Here, a JSContext is set up at ❶, an instance of a User class is set up at ❷, and some values are assigned to three of the new user's properties. One of those properties, hiddenName, was defined only in the @implementation instead of the protocol—the same thing that happened in Listing 9-6 with the password property. At ❸, the newly created user is bridged to the JSContext. When the code subsequently tries to access the values of the user object from JavaScript, the value property is successfully accessed at ❹, while the attempt to access hiddenName fails ❺.

NOTE   *Use discretion when exporting objects to JavaScriptCore. An attacker who exploits a script injection flaw will be able to run any exported functions, essentially turning the script injection into native remote code execution on users' devices.*

One additional interesting point is that JavaScriptCore disallows calling exported class constructors. (This is a bug in iOS that, as of iOS 8, has yet to be resolved.) So even if you add [User class] to your context, you won't be able to create new objects using new. As I discovered through some testing, however, it's possible to work around that limitation. You can essentially implement an exported Objective-C block that accepts a class name and then creates and returns an instance of an arbitrary class to JavaScript, as I've done here:

```
self.context[@"newInstance"] = ^(NSString *className) {
    Class clazz = NSClassFromString(className);
    id inst = [clazz alloc];
    return inst;
};

[self.context evaluateScript:@"var u = newInstance('User');"];
JSValue *val = self.context[@"u"];
User *user = [val toObject];
NSLog(@"%@", [user class]); // => User
```

This technique bypasses the need to explicitly export any classes and allows you to instantiate an object of any type and expose it to JavaScript-Core. However, no members have been whitelisted to be exported, so there are no strong references to any methods or variables of the class object. Clearly, there's plenty of room for more security research into bypassing the restrictions implemented by JavaScriptCore because the Objective-C runtime is such a dynamic and powerful beast.

One common complaint about the JavaScriptCore framework is that there is no documented way to access the `JSContext` of a `UIWebView`. I'll discuss some potential ways around this next.

Manipulating JavaScript in Web Views

Why expose this `JSContext` functionality without a way to access it within a web view? It's not clear what Apple's intentions were, but the developers did only half the job of documenting the JavaScriptCore APIs. As of yet, there's no official Apple way to manipulate a `UIWebView`'s `JSContext`, but several people have discovered methods to do so. Most of them involve using the `valueForKeyPath` method, as in Listing 9-7.

```
- (void)webViewDidFinishLoad:(UIWebView *)webView {
    JSContext *context = [webView valueForKeyPath:@"documentView.webView.
  mainFrame.javaScriptContext"];

    context[@"document"][@"cookie"] = @"hello, I'm messing with cookies";
}
```

Listing 9-7: Manipulating a DOM via Objective-C

Since this isn't an officially Apple-sanctioned approach, there's no guarantee that this kind of code will make it into the App Store, but it's worth being aware of the ways developers may try to communicate between JavaScript and Objective-C and the pitfalls it poses.

Of course, the `JSContext` isn't the only way to connect JavaScript to Objective-C. I describe Cordova, another popular bridge, in the next section.

Executing JavaScript with Cordova

Cordova (known as PhoneGap before Adobe acquired the development firm Nitobi) is an SDK that provides native mobile APIs to a web view's JavaScript execution environment in a platform-agnostic manner. This allows mobile applications to be developed like standard web applications using HTML, CSS, and JavaScript. Those applications then work across all platforms Cordova supports. This can significantly reduce lead time and do away with the need for development firms to hire platform-specific engineers, but Cordova's implementation increases the application attack surface significantly.

How Cordova Works

Cordova bridges JavaScript and Objective-C by implementing an NSURLProtocol to handle any JavaScript-initiated XmlHttpRequest to *file://!gap_exec*. If the native Cordova library detects a call to this URI, it attempts to pull class, method, argument, and callback information out of the request headers, as evidenced in Listing 9-8.

```
+ (BOOL)canInitWithRequest:(NSURLRequest*)theRequest {
    NSURL* theUrl = [theRequest URL];
    CDVViewController* viewController = viewControllerForRequest(theRequest);

    if ([[theUrl absoluteString] hasPrefix:kCDVAssetsLibraryPrefixs]) {
        return YES;
    } else if (viewController != nil) {
❶      if ([[theUrl path] isEqualToString:@"/!gap_exec"]) {
❷          NSString* queuedCommandsJSON = [theRequest valueForHTTPHeaderField:@"
    cmds"];
            NSString* requestId = [theRequest valueForHTTPHeaderField:@"rc"];
            if (requestId == nil) {
                NSLog(@"!cordova request missing rc header");
                return NO;
            }
            BOOL hasCmds = [queuedCommandsJSON length] > 0;
            if (hasCmds) {
                SEL sel = @selector(enqueCommandBatch:);
❸              [viewController.commandQueue performSelectorOnMainThread:sel
    withObject:queuedCommandsJSON waitUntilDone:NO];
```

Listing 9-8: Detecting native library calls in CDVURLProtocol.m[3]

At ❶, the request URL is checked for a path component of */!gap_exec*, and at ❷, the value of the cmds HTTP header is extracted. Cordova then passes these commands to the command queue ❸, where they will be executed if possible. When these commands are queued, Cordova looks up the information in a map of available Cordova plug-ins, which essentially just expose various portions of the native functionality and can be extended arbitrarily. If a particular plug-in is enabled and the class in the request can be instantiated, then the method is called with the supplied arguments using the all-powerful objc_msgSend.

When the call completes, the native code calls back to the JavaScript runtime via [UIWebView stringByEvaluatingJavaScriptFromString], calling the cordova.require('cordova/exec').nativeCallback method defined in *cordova.js*, and provides the original callback ID as well as the return value of the native code execution.

---

3. *https://github.com/apache/cordova-ios/blob/master/CordovaLib/Classes/Public/CDVURLProtocol.m*

This exports an unprecedented amount of native object control to the JavaScript runtime, allowing applications to read and write files, read and write Keychain storage, upload local files to a remote server via FTP, and so on. But with this increased functionality comes potential pitfalls.

Risks of Using Cordova

If your app contains any script injection vulnerabilities and if your users can influence application navigation, an attacker could obtain remote code execution. They would just have to inject callback functions combined with a call to initiate communication with native code. For instance, an attacker might inject a call to access Keychain items, grab a copy of all the user's contacts, or read out a file and feed it into a JavaScript function of their choosing, as demonstrated in Listing 9-9.

```
<script type="text/javascript">
    var exec = cordova.require('cordova/exec');
    function callback(msg) {
        console.log(msg);
    }
    exec(callback, callback, "File", "readAsText", ["/private/var/mobile/Library/
     Preferences/com.apple.MobileSMS.plist", "UTF-8",
        0, 2048]);
</script>
```

Listing 9-9: Using Cordova to make Objective-C calls to read the contents of a file

This attacker-supplied JavaScript reads the device's *com.apple.MobileSMS .plist*, which, in iOS 8, is accessible to all applications on the device.[4] This gives the attacker the ability to examine the user's contacts, as well as determine the owner of the device in question.

One reasonable bit of built-in security that can significantly reduce the risks of script injection is *domain whitelisting*.[5] Cordova's default security policy blocks all network access and allows interaction only with domains that are whitelisted under the <access> element in the app configuration. The whitelist does allow access to all domains via a wildcard (*) entry, but don't be lazy—ensure that only the domains your app needs to talk to in order to function properly are in the whitelist. You can configure this through Xcode by adding values to the `ExternalHosts` key in `Cordova.plist`, as shown in Figure 9-1.

---

4. *http://www.andreas-kurtz.de/2014/09/malicious-apps-ios8.html*

5. *http://docs.phonegap.com/en/1.9.0/guide_whitelist_index.md.html*

Key		Type	Value
▼ Root		Dictionary	(12 items)
UIWebViewBounce		Boolean	YES
TopActivityIndicator		String	gray
EnableLocation		Boolean	NO
EnableViewportScale		Boolean	NO
AutoHideSplashScreen		Boolean	YES
ShowSplashScreenSpinner		Boolean	YES
MediaPlaybackRequiresUserAction		Boolean	NO
AllowInlineMediaPlayback		Boolean	NO
OpenAllWhitelistURLsInWebView		Boolean	NO
BackupWebStorage		Boolean	YES
▼ ExternalHosts	⊕ ⊖	Array	(1 item)
Item 0	⊕ ⊖	String	isecpartners.com
▼ Plugins		Dictionary	(17 items)
Device		String	CDVDevice
Logger		String	CDVLogger
Compass		String	CDVLocation
Accelerometer		String	CDVAccelerometer
Camera		String	CDVCamera
NetworkStatus		String	CDVConnection
Contacts		String	CDVContacts

Figure 9-1: Whitelisting domains in Cordova using the `ExternalHosts` *key*

Besides exposing native code objects to the web view, there are many other drawbacks to implementing mobile applications using a web platform wrapper such as Cordova. Mainly, each mobile platform has its own security model predicated on specific assumptions, APIs, and functionality to protect users and secure local storage. One platform's security model just won't make sense on other platforms. Providing a one-size-fits-all implementation is, necessarily, going to exclude some of these platform-specific security benefits for the sake of usability.

For example, iOS provides secure storage through the Data Protection APIs (as I describe in Chapter 13), which require specific arguments that don't lend themselves to a cross-platform implementation. As such, these APIs are not supported by Cordova, preventing fine-grained control over when file data is encrypted at rest. To solve this problem, you can enable entitlement-level data protection (refer to "The DataProtectionClass Entitlement" on page 223), which will apply a default protection level ubiquitously for all data written to disk by the application.

Another common issue is the lack of a similar secure storage element across platforms. This removes direct Keychain access on iOS, although Adobe ultimately developed an open source plug-in[6] to address the problem.

That ends the tour of `UIWebView` and JavaScript bridges, but new applications (for iOS 8 and newer) will increasingly be using the `WKWebView` API. I'll cover how to wrangle `WKWebView` in the following section.

---

6. *https://github.com/shazron/KeychainPlugin*

Enter WKWebView

As I mentioned previously, a newer interface to WebKit was introduced with iOS 8 to supplant `UIWebView`. `WKWebView` addresses several of the shortcomings of `UIWebView`, including access to the Nitro JavaScript engine, which greatly increases performance on JavaScript-heavy tasks. Let's look at how apps would create `WKWebViews` and how `WKWebViews` can improve your app's security.

Working with WKWebViews

A `WKWebView` is instantiated in essentially the same way as a `UIWebView`, as shown here:

```
CGRect webFrame = CGRectMake(0, 0, width, height);
WKWebViewConfiguration *conf = [[WKWebViewConfiguration alloc] init];
WKWebView *webView =[[WKWebView alloc] initWithFrame:webFrame
                                       configuration:conf];
NSURL *url = [NSURL URLWithString:@"http://www.nostarch.com"];
NSURLRequest *request = [NSURLRequest requestWithURL:url];
[webView loadRequest:request];
```

This just allocates a new `WKWebView` instance and then initializes it with the `initWithFrame` method.

To customize behavior, `WKWebViews` can also be instantiated with user-supplied JavaScript, as in Listing 9-10. This allows you to load a third-party website but with your own custom JavaScript that executes upon page load.

```
  CGRect webFrame = CGRectMake(0, 0, width, height);
❶ NSString *src = @"alert('Welcome to my WKWebView!')";
❷ WKWebViewConfiguration *conf = [[WKWebViewConfiguration alloc] init];
❸ WKUserScript *script = [[WKUserScript alloc] initWithSource:src
          injectionTime:WKUserScriptInjectionTimeAtDocumentStart
        forMainFrameOnly:YES];
❹ WKUserContentController *controller = [[WKUserContentController alloc] init];
❺ [conf setUserContentController:controller];
❻ [controller addUserScript:script];
❼ WKWebView *webView =[[WKWebView alloc] initWithFrame:webFrame
                                         configuration:conf];
```

Listing 9-10: Instantiating a `WKWebView` *with custom JavaScript*

At ❶, a simple `NSString` that consists of a single JavaScript command is created. At ❷, a configuration object is created that will hold the configuration parameters for the web view that will be created later. At ❸, a `WKUserScript` object is created and initialized with the src that contains the JavaScript you want to execute. Then a `WKUserContentController` is made at ❹, which is set in the configuration object at ❺. Finally, the script is added to the controller with the addUserScript method at ❻, and the web view is instantiated at ❼.

NOTE *As with other methods of injecting JavaScript, be careful not to interpolate content provided by third parties without strict sanitization.*

Security Benefits of WKWebViews

Using WKWebViews has a couple security advantages. First, you can set preferences that disable loading JavaScript with the method setJavaScriptEnabled if the pages you plan to load don't require it; if the remote site has malicious script, this will prevent that script from executing. You can also leave JavaScript enabled but disable the opening of new windows from JavaScript using the setJavaScriptCanOpenWindowsAutomatically method—this will prevent most pop-ups from opening, which can be quite irritating in web views.

Lastly, and perhaps most importantly, you can actually detect whether the contents of the web view were loaded over HTTPS, giving you the ability to ensure that no parts of the page were loaded over insecure channels. For UIWebViews, there is no indication to the user or developer when the web view loads mixed content—the hasOnlySecureContent method of WKWebView resolves this problem. Listing 9-11 shows a way to implement a somewhat hardened WKWebView.

```
@interface ViewController ()
@property (strong, nonatomic) WKWebView *webView;

@end

@implementation ViewController

- (void)viewDidLoad {
    [super viewDidLoad];

❶  WKPreferences *pref = [[WKPreferences alloc] init];
    [pref setJavaScriptEnabled:NO];
    [pref setJavaScriptCanOpenWindowsAutomatically:NO];

❷  WKWebViewConfiguration *conf = [[WKWebViewConfiguration alloc] init];
    [conf setPreferences:pref];

❸  NSURL *myURL = [NSURL URLWithString:@"https://people.mozilla.org/~mkelly/
    mixed_test.html"];

❹  _webView = [[WKWebView alloc] initWithFrame:[[self view] frame]
                                configuration:conf];
    [_webView setNavigationDelegate:self];
❺  [_webView loadRequest:[NSURLRequest requestWithURL:myURL]];
```

```
        [[self view] addSubview:_webView];
    }

❻  - (void)webView:(WKWebView *)webView didFinishNavigation:(WKNavigation *)navigation
    {
        if (![webView hasOnlySecureContent]) {

            NSString *title = @"Ack! Mixed content!";
            NSString *message = @"Not all content on this page was loaded securely.";
            UIAlertView *alert = [[UIAlertView alloc] initWithTitle:title
                                                            message:message
                                                           delegate:self
                                                  cancelButtonTitle:@"OK"
                                                  otherButtonTitles:nil];

            [alert show];
        }
    }
```

Listing 9-11: A nice, safe WKWebView

This code uses a couple of extra security mechanisms that WKWebView provides. At ❶, a WKPreferences instance is instantiated, and the setJavaScript-Enabled and setJavaScriptCanOpenWindowsAutomatically properties are set on it. (These are redundant, of course, but you can pick whichever property best suits your needs.) Then, a WKWebViewConfiguration object is instantiated at ❷ and the WKPreferences already created are passed in. At ❸, a URL to load is defined; in this case, it's simply an example page that includes mixed content. At ❹, the WKWebView itself is made, using the configuration created previously. The code then requests that the web view load a given URL at ❺. Finally, the didFinishNavigation delegate is implemented at ❻, which in turn calls hasOnlySecureContent on the web view. If the content is mixed, the user is alerted.

Closing Thoughts

While modern versions of iOS have made great strides in allowing developers control over the interactions between native code and web content, there is a legacy of hacks to bridge the two, with their own idiosyncrasies. At this point, you should be aware of the main bridging mechanisms, as well as where to look for potentially malicious externally supplied data.

I also briefly covered some of the caching that takes place when working with web content. In Chapter 10, you'll dig in to the many ways that data can leak to the local filesystem and be recovered by attackers.

10

DATA LEAKAGE

Data theft is a serious concern in the mobile world, where devices containing critical personal and business data are lost or stolen frequently. The primary threat to consider here is forensic attackers, so use special care to ensure that such data is persisted in a format that can't be easily extracted by physical attackers or by compromised devices. Unfortunately, there's a lot of confusion over what APIs actually end up storing sensitive data, which is understandable since much of this behavior is undocumented.

In this chapter, I'll examine the many ways in which data can leak from your application to dark corners of the device—and even accidentally be synced to remote services such as iCloud. You'll learn how to search for leaked data on a device or within your own Simulator application directory structure, as well as how to prevent these kinds of leaks from happening.

The Truth About NSLog and the Apple System Log

For years developers have used `printf` to output basic debug information while writing programs. In iOS, *NSLog* appears to be the equivalent, and it's frequently used as such. However, `NSLog` doesn't merely write output to the

Xcode console, as most people believe. Its purpose is to log an error message to the Apple System Log (ASL) facility. Here's what Apple has to say:

> Messages received by the server are saved in a data store (subject to input filtering constraints). This API permits clients to create queries and search the message data store for matching messages.[1]

So perhaps NSLog is best thought of as a hybrid between printf and syslog, which spits out messages in the Xcode console when debugging and sends messages to a global system log when on the device. It follows, then, that data logged by NSLog will be retrievable by anyone in physical possession of the device, similar to other cached application data.

No special tools are necessary to read the log. Just plug the iOS device in to a Mac, open Xcode, select **Window** → **Devices**, and click your device. The device's system log may not be initially visible in the console. If it isn't, click the tiny arrow in the lower left of the panel. Figure 10-1 shows an example of viewing the console log with the Devices window.

Figure 10-1: The Devices window in Xcode

The Apple System Log facility has one quirk that makes it different from the traditional UNIX syslog facility: you can create queries to search existing data in the ASL. In versions of iOS before iOS 7, this function works regardless of which application originally submitted the data, which means that

---

1. *https://developer.apple.com/library/ios/#documentation/System/Conceptual/ManPages_iPhoneOS/man3/asl.3.html*

any information an application logs can be read by any other application on the device. Any application can read the ASL programmatically, too, as Oliver Drobnik describes on the Cocoanetics blog.[2] In fact, there are several applications that act as system log viewers using this API.

In iOS 7 and later, the impact of this flaw has lessened significantly because apps can access only their own logs. However, all application logs can still be read with physical access to a device, provided that the device has a trust relationship with another computer (or that the attacker jailbreaks the device).

Since log information can leak under certain circumstances, you need to be painstakingly careful to ensure that sensitive information doesn't end up in the system log. For example, I've seen applications containing code like the horrible snippet in Listing 10-1.

```
NSLog(@"Sending username \%@ and password \%@", myName, myPass);
```

Listing 10-1: Please don't do this.

If you're sending usernames, passwords, and so on, to NSLog, you're basically handing over users' private information, and you should feel bad about that. To redeem yourself, stop abusing NSLog; take it out of the equation before releasing your app to users.

Disabling NSLog in Release Builds

The simplest way to get rid of NSLog output is to use a variadic macro (Listing 10-2) that makes NSLog a no-op unless the app is built in Debug mode within Xcode.

```
#ifdef DEBUG
#    define NSLog(...) NSLog(__VA_ARGS__);
#else
#    define NSLog(...)
#endif
```

Listing 10-2: Disabling NSLog in nondebug builds

As bad as NSLog seems, apps with NSLog *do* make it into the App Store. This may change at some point, but you can't rely on Apple to detect that your application is logging information that you don't intend, nor can you rely on Apple to prevent applications from reading that logged data.

---

2. *http://www.cocoanetics.com/2011/03/accessing-the-ios-system-log/*

Logging with Breakpoint Actions Instead

Another option is to use breakpoint actions to do logging, as I touched on in Chapter 5. In that case, you're effectively logging with the debugger, rather than the program itself. This is more convenient in some circumstances and doesn't result in data being written to the system log when deployed, reducing the risk of releasing code with logging enabled to zero. Knowing how to use these actions will also be useful to you in future debugging.

Breakpoint actions are stored within a project, rather than in the source itself. They're also user specific, so you see only the breakpoints and logging actions that you care about, rather than having everyone on your team clutter up the codebase with their logging statements. But when needed, Xcode lets you share your breakpoints with other users, making them part of the main project (see Figure 10-2).

You can also easily enable or disable actions, as well as specify that they shouldn't output until the breakpoint is hit a certain number of times. You can even specify complex breakpoint conditions, which define when the associated actions will execute.

If you want to disable all the breakpoints in a project, you can do this a couple of ways in Xcode. Either go the breakpoint navigator and right-click the workspace icon (Figure 10-2) or use the shortcut ⌘-Y.

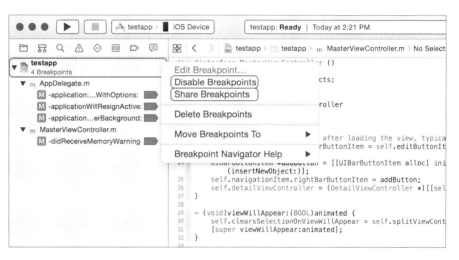

Figure 10-2: Sharing breakpoints with other users and disabling all breakpoints in Xcode

While NSLog leaks information to disk where it can be read by a physical attacker (and malicious apps in some versions of iOS), data can also leak between apps via more transient mechanisms, such as iOS pasteboards. Let's take a look at them now.

How Sensitive Data Leaks Through Pasteboards

The iOS pasteboard is a flexible mechanism for sharing arbitrary data within or between applications. Via a pasteboard, you can share textual data or serialized objects between applications, with the option to persist these pasteboards on disk.

Restriction-Free System Pasteboards

There are two default system pasteboards: `UIPasteboardNameGeneral` and `UIPasteboardNameFind`. The former is the pasteboard that almost every application will read from and write to by default when using Cut, Copy, or Paste menu items from within the app, and it's the pasteboard of choice when you want to share data between third-party applications. The latter is a special pasteboard that stores the contents of the last search string entered into a `UISearchBar`, so applications can automatically determine what users have searched for in other applications.

NOTE *Contrary to the official description of `UIPasteboardNameFind`, this pasteboard is never used in real life. This bug is acknowledged by Apple but hasn't been fixed, nor has the documentation been updated. As a security consultant, I can only hope that it will be fixed so that I can complain about it being a security flaw.*

It's important to remember that the system pasteboards have *no* access controls or restrictions. If your application stores something on the pasteboard, any application has access to read, delete, or tamper with that data. This tampering can come from processes running in the background, polling pasteboard contents periodically to harvest sensitive data (see Osamu Noguchi's UIPasteBoardSniffer[3] for a demonstration of this technique). As such, you need to be extremely careful about what ends up on `UIPasteboardNameGeneral` in particular, as well as pasteboards in general.

The Risks of Custom-Named Pasteboards

Custom-named pasteboards are sometimes referred to as *private* pasteboards, which is an unfortunate misnomer. While applications can create their own pasteboards for internal use or to share among other specific applications, custom pasteboards are *public* in versions of iOS prior to 7, making them available for any program to use so long as their names are known.

Custom pasteboards are created with `pasteboardWithName`, and in iOS 7 and later, both `pasteboardWithName` and `pasteboardWithUniqueName` are specific to all applications within an application group. If other applications outside of this group attempt to create a pasteboard with a name already in use, they'll be assigned a totally separate pasteboard. Note, however, that the two system pasteboards are still accessible by any application. Given that a number of devices can't be upgraded to iOS 6, much less iOS 7, you should carefully examine how custom pasteboards are used in different versions of iOS.

One thing that you can do with a custom pasteboard is mark it as persistent across reboots by setting the `persistent` property to `YES`. This will cause pasteboard contents to be written to *$SIMPATH/Devices/<device ID>/*

---

3. *https://github.com/Atrac613/UIPasteboardSniffer-iOS*

data/Library/Caches/com.apple.UIKit.pboard/pasteboardDB, along with other
application pasteboards. Listing 10-3 shows some data you might see in the
pasteboardDB file.

```
<?xml version="1.0" encoding="UTF-8"?>
<!DOCTYPE plist PUBLIC "-//Apple//DTD PLIST 1.0//EN" "http://www.apple.com/DTDs/
    PropertyList-1.0.dtd">
<plist version="1.0">
<array>
        <integer>1</integer>
        <dict>
                <key>bundle</key>
                <string>com.apple.UIKit.pboard</string>
                <key>items</key>
                <array/>
                <key>itemsunderlock</key>
                <array/>
                <key>name</key>
                <string>com.apple.UIKit.pboard.find</string>
                <key>persistent</key>
                <true/>
        </dict>

    --snip--

        <dict>
                <key>bundle</key>
                <string>com.apple.UIKit.pboard</string>
                <key>items</key>
                <array>
                        <dict>
                                <key>Apple Web Archive pasteboard type</key>
                                <data>
                                bigbase64encodedblob==
                                </data>
                                <key>public.text</key>
                                <data>
                                aHR0cDovL2J1cnAvY2VydA==
                                </data>
                        </dict>
                </array>
                <key>itemsunderlock</key>
                <array/>
                <key>name</key>
                <string>com.apple.UIKit.pboard.general</string>
                <key>persistent</key>
                <true/>
```

```
        </dict>
    </array>
</plist>
```

Listing 10-3: Possible contents of the com.apple.UIKit.pboard/pasteboardDB *file*

The base64 blobs `bigbase64encodedblob` (too big to include in its entirety) and `aHR0cDovL2J1cnAvY2VydA` hold pasteboard contents, leaving those contents accessible to any application that can read *pasteboardDB*. Note, too, that pasteboards can be of different types: the Apple Web Archive pasteboard allows an entire web page to be stored, while the `public.text` pasteboard is the text content of the general pasteboard.[4]

Pasteboard Data Protection Strategies

To minimize the risk of information leakage, it's a good idea to analyze exactly what behavior you're trying to facilitate by using pasteboards. Here are some questions to ask yourself:

• Do I want users to copy information into other applications, or will they simply need to move data within my application?

• How long should clipboard data live?

• Is there any place in the application that data should never be copied from?

• Is there any part of the application that should never receive pasted data?

The answers to these questions will inform the way you should handle pasteboard data within your application. You can take a few different approaches to minimize data leakage.

Wiping the Pasteboard When You Switch Apps

If you want your users to copy and paste only within your own application, you can clear the pasteboard on the appropriate events to ensure that data doesn't stay on the pasteboard when the user switches applications. To do this, clear the pasteboard by setting `pasteBoard.items = nil` on the `applicationDidEnterBackground` and `applicationWillTerminate` events. This won't prevent applications running in the background from reading the pasteboard, but it will shorten the lifetime of the data on the pasteboard and will prevent users from pasting data into apps they're not supposed to.

Keep in mind that clearing the pasteboard may interfere with data the end user or other applications are using for a different purpose. You may want to create a flag that indicates whether potentially sensitive data has been written to the pasteboard and clear it only conditionally.

---

4. *https://developer.apple.com/library/mac/documentation/Cocoa/Reference/WebKit/Classes/WebArchive_Class/Reference/Reference.html*

Preventing Copy/Paste Selectively

Even when you do want to let the user copy and paste, sometimes there are specific places you want to disallow the option. For example, you might want to prevent a user from pasting in a PIN or answer to a security question (such data should never be on the pasteboard in the first place) yet allow the ability to paste in an email address from an email.

NOTE *That's not to say you should use security questions, which tend to enable account hijacking by using publicly available information as an authenticator. You'll take a look at this in "Keylogging and the Autocorrection Database" on page 175.*

The official way to allow users to paste some information and prevent them from pasting others is with the canPerformAction:withSender responder method.[5] Create a new class in Xcode, as in Figure 10-3.

Figure 10-3: Creating the restrictedUITextField *subclass*

Then, edit *restrictedUITextField.m* and add the canPerformAction method.

```
#import "restrictedUITextField.h"

@implementation restrictedUITextField

- (id)initWithFrame:(CGRect)frame {
    self = [super initWithFrame:frame];
    if (self) {
        // Initialization code
    }
    return self;
}

❶ -(BOOL)canPerformAction:(SEL)action withSender:(id)sender {
❷     if (action == @selector(cut:) || action == @selector(copy:))
        return NO;
    else
        return YES;
}
@end
```

Listing 10-4: Adding canPerformAction *to* restrictedUITextField.m

5. *http://developer.apple.com/library/ios/#documentation/uikit/reference/UIResponder_Class/Reference/Reference.html*

The `canPerformAction` method at ❶ is passed an `action` selector, which can be inspected to see what type of action is being requested at ❷. You can use any method from those specified in the `UIResponderStandardEditActions` protocol. If you want to entirely disable the context menu, you can, of course, simply return `NO` in every circumstance.

Finding and Plugging HTTP Cache Leaks

You'll also find cached data from the URL loading system stored, unencrypted, in the *<app ID>/Library/Caches/com.mycompany.myapp/Cache.db** files, which are private to each application. HTTP caches can contain images, URLs, and text fetched over HTTP and can therefore expose sensitive data to a third party if examined. An easy way to get an idea of the type of data exposed by your application is to use File Juicer to carve the database into normal, readable individual files. You can download File Juicer at *http://echoone.com/filejuicer/*, and Figure 10-4 shows the type of output it provides.

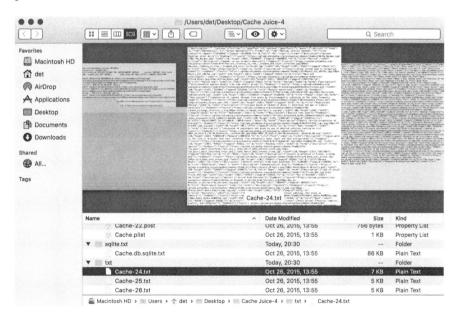

Figure 10-4: Examining the contents of the cache databases, split into separate files and directories by File Juicer

File Juicer splits data into directories based on particular file types, so you can investigate stored images, plists, SQLite databases, or plaintext conversions of other binary file types.

Once you know what kind of data your application exposes through cached data, you can consider how best to manage it.

Cache Management

Cache management on iOS is somewhat complex. There are many configuration settings and a seemingly endless number of ways to affect cache policy. On top of that, the platform tries to aggressively cache and copy everything it can get its hands on to try to improve the user experience. Developers need to determine which of these methods allows for secure cache management, but it's easy to lull yourself into a false sense of security. Pentesters have to know when clients who think they are doing the right things are in fact leaking potentially sensitive information onto disk. Let's talk about all the wrong ways to manage caches.

As I mentioned in Chapter 5, the documented way to remove cached data, [NSURLCache removeAllCachedResponses], only removes cache entries from memory. This is essentially useless for security purposes because the same information is persisted to disk and is not removed. Perhaps there's a better approach.

Ideally, you won't ever need to delete the cache because removal implies that you were caching responses in the first place. If the response data is so sensitive, then why not just never cache it? Let's give that a shot.

The first place to start limiting cached responses is in the NSURLCache configuration, as in Listing 10-5. This API lets you control the amount of memory and disk capacity that the platform dedicates to the cache.

```
NSURLCache *urlCache = [[NSURLCache alloc] init];
[urlCache setDiskCapacity:0];
[NSURLCache setSharedURLCache:urlCache];
```

Listing 10-5: Limiting disk cache storage to zero bytes

The problem with this strategy is that the capacity manipulation APIs are not intended to be security mechanisms. Rather, these configurations exist to provide the system with information to be used when memory or disk space runs low. The NSURLCache documentation[6] specifies that both the on-disk and in-memory caches will be truncated to the configured sizes only if necessary.

So you can't trust configuring the cache capacity. What about setting the cache policy to NSURLRequestReloadIgnoringLocalCacheData to force the URL loading system to ignore any cached responses and fetch the data anew? Here's how that might work:

```
NSURLRequest* req = [NSURLRequest requestWithURL:aURL
            cachePolicy:NSURLRequestReloadIgnoringLocalCacheData
        timeoutInterval:666.0];
  [myWebView loadRequest:req];
```

---

6. *https://developer.apple.com/library/ios/documentation/cocoa/reference/foundation/Classes/ NSURLCache_Class/Reference/Reference.html#//apple_ref/occ/instm/NSURLCache/setDiskCapacity:*

But this policy is not implicitly preventing responses from being cached; it's preventing the URL loading system only from *retrieving* the cached responses on subsequent fetches. Any previously cached responses will persist on disk, which poses problems if your initial app implementations allowed caching. No dice.

As I've tried to demonstrate, if you rely on the system defaults for web view cache management, you might just implement a lot of precautions that don't really protect users at all. If you want to reliably control the contents of your application caches, you need to do it yourself. Luckily, this isn't actually that difficult.

The Cocoa Touch API gives developers the ability to manipulate responses on a per-request basis before they are cached using the [NSURLConnection connection:willCacheResponse:] method. If you don't want to cache the data, you can implement the delegate method, as shown in Listing 10-6.

```
-(NSCachedURLResponse *)connection:(NSURLConnection *)connection
                willCacheResponse:(NSCachedURLResponse *)cachedResponse {
  NSCachedURLResponse *newCachedResponse = cachedResponse;
  if ([[[[cachedResponse response] URL] scheme] isEqual:@"https"]) {
    newCachedResponse=nil;
  }
  return newCachedResponse;
}
```

Listing 10-6: Preventing caching of responses served over secure connections

This implementation of the delegate just returns NULL instead of the NSCachedURLResponse representation of the response data.

Similarly, for data fetched using the NSURLSession class, you'd implement the [NSURLSessionDataDelegate URLSession:dataTask:willCacheResponse:completion-Handler:] delegate method. Beware of relying entirely on this method, however, because it is called only for data and upload tasks. Caching behavior for download tasks will still be determined by the cache policy only and should be resolved similarly to Listing 10-6.[7]

In summary, caching on iOS is unreliable. Be careful, and double-check your app after extended use to make sure it's not leaving sensitive information around.

Solutions for Removing Cached Data

The documented way to remove locally cached data is to use the removeAllCachedResponses method of the shared URL cache, shown in Listing 10-7.

---

7. *https://developer.apple.com/library/ios/documentation/Cocoa/Conceptual/URLLoadingSystem/ Concepts/CachePolicies.html#//apple_ref/doc/uid/20001843-BAJEAIEE*

```
[[NSURLCache sharedURLCache] removeAllCachedResponses];
```

Listing 10-7: The documented API for removing cached data

A similar method, removeCachedResponseForRequest, exists to remove cached data for only specific sites. However, as you discovered in Chapter 4, this removes only cached data from memory and not from the disk cache that you're actually concerned with. I would file a bug, if Apple's bug tracking system were not an infinitely hot and dense dot from which no light or information could escape.[8] Anyway, there are a few ways you can work around this—the caching issue, I mean; you're on your own if you're unfortunate enough to have to report a bug.

Just Don't Cache

In most circumstances, it's better to just prevent caching altogether, rather than clean up piecemeal afterward. You can proactively set the cache capacities for disk and memory to zero (Listing 10-8), or you can simply disable caching for the disk, if you're comfortable with in-RAM caching.

```
- (void)applicationDidFinishLaunching:(UIApplication *)application {
    [[NSURLCache sharedURLCache] setDiskCapacity:0];
    [[NSURLCache sharedURLCache] setMemoryCapacity:0];
    // other init code
}
```

Listing 10-8: Disallowing cache storage by limiting permitted storage space

Alternatively, you can implement a willCacheResponse delegate method of NSURLConnection, returning a value of nil, as in Listing 10-9.

```
-(NSCachedURLResponse *)connection:(NSURLConnection *)connection
                willCacheResponse:(NSCachedURLResponse *)cachedResponse {
    NSCachedURLResponse *newCachedResponse=cachedResponse;
❶   if ([cachedResponse response]) {
❷       newCachedResponse=nil;
    }
  return newCachedResponse;
}
```

Listing 10-9: Sample cache discarding code

---

8. It's rather out of character for me to not file bugs, but Apple's bug tracker, RADAR, is so breathtakingly, insultingly useless that no reasonable person should have to use it. Instead, I recommend visiting *http://fixradarorgtfo.com/* and filing this single RADAR bug: "Fix Radar or GTFO (duplicate of rdar://10993759)."

This just checks whether a cached response has been sent at ❶ and, if it finds one, sets it to `nil` at ❷. You can also conditionally cache data by examining the properties of the response before returning the object to cache, as shown in Listing 10-10.

```
-(NSCachedURLResponse *)connection:(NSURLConnection *)connection
                willCacheResponse:(NSCachedURLResponse *)cachedResponse {
    NSCachedURLResponse *newCachedResponse=cachedResponse;
    if ([[[[cachedResponse response] URL] scheme] isEqual:@"https"]) {
        newCachedResponse=nil;
    }
  return newCachedResponse;
}
```

❶ appears beside the `if` line.

Listing 10-10: Conditional cache discarding code

This is nearly the same as in Listing 10-9, but it additionally examines the response being cached at ❶ to determine whether it is being delivered over HTTPS and discards it if that's the case.

If you're using `NSURLSession`, you can also use ephemeral sessions, which will not store any data to disk; this includes caches, credentials, and so forth. Creating an ephemeral session is easy. Just instantiate a configuration object for your `NSURLSessions`, like so:

```
NSURLSessionConfiguration *config = [NSURLSessionConfiguration
    ephemeralSessionConfiguration];
```

You can find more information and examples of how to use `NSURLSession` in Chapter 7.

Disable Caching via the Server

Presuming you control the server your application communicates with, you can instruct the client not to cache requests using the `Cache-Control` HTTP header. This allows you to either disable caching application-wide or apply it only to specific requests. The mechanism for implementing this on the server side is language-dependent, but the header you'll want to return for requests you don't want cached is as follows:

```
Cache-Control: no-cache, no-store, must-revalidate
```

Sadly, at least some versions of iOS (verified as of 6.1) don't actually obey these headers. It's a good idea to set them for sensitive resources regardless, but don't rely on this method to solve the problem entirely.

Go Nuclear

The previous approaches will prevent data from being cached, but sometimes you may want to cache data and then clean it up later. This could be
for performance reasons or perhaps because you're correcting a caching
problem in a newly released version of your application, which has already
cached data locally on disk. Whatever your reason, clearing the cache in the
documented manner doesn't work, so you're stuck removing the cached
data forcibly, as in Listing 10-11.

```
NSString *cacheDir=[NSSearchPathForDirectoriesInDomains(NSCachesDirectory,
    NSUserDomainMask, YES) objectAtIndex:0];


[[NSFileManager defaultManager] removeItemAtPath:cacheDir
                                    error:nil];
```

Listing 10-11: Manually removing the cache database

There's no guarantee that some other part of the system won't freak out
if you clear cached data manually. However, this method is the only reliable
way I've found to remove cached data after it's already been written to disk.

Data Leakage from HTTP Local Storage and Databases

The HTML 5 specification allows websites to store and retrieve large amounts
of data (larger than what would fit in a cookie) on the client. These mechanisms are sometimes used to cache data locally so that primarily web-based
applications can function in an offline mode. You can find these databases
in a number of locations on the device or your simulator, including the
following:

- */Library/Caches/*.localstorage*
- */Library/Caches/Databases.db*
- */Library/Caches/file__0/*.db*

You can feed these locations to File Juicer the same way you do with
HTTP caches to get access to the plaintext data. One obvious application
for larger local storage and SQL databases is storing structured information
about communications such as email so that those communications can be
accessed when the user doesn't have cell phone reception. This can leave
traces around the storage databases, as shown in Figure 10-5.

This exposure is probably an acceptable risk for metadata, but storing
it in an encrypted SQLite store might be better, especially when storing full
message contents. I'll talk more about how to do this in "Data Protection
API" on page 219.

```
SQLite format 3   @
tableItemTableItemTableCREATE TABLE ItemTable TEXT UNIQUE ON CONFLICT  REPLACE,  value
BLOB  NOT  NULL  ON  CONFLICT
sp/hybrid/bobrobertson666@gmail.com/imp-ifÅ
pref/bobrobertson666@gmail.com/asfalse^Å- pref/bobrobertson666@gmail.com/bx_asnsfalseR
pref/bobrobertson666@gmail.com/cstrueR" pref/bobrobertson666@gmail.com/ednull\#Å
Robertson",true]]R&Åpref/bobrobertson666@gmail.com/pitrueÅ?'Å
Robertson",true]X(Å
pref/bobrobertson175@gmail.com/kbsfalse`?Åsp/hybrid/bobrobertsþn666@gmail.com/db-
rcc43h9Å5,sp/hybrid/bobrobertson666@gmail.com/aq-l{"ca":0}◊;
(label:^iim)) (((!label:^s) (!label:^k)) ((label:^smartlabel_personal) OR  (label:^t)))  OR
(((label:^i)  OR  (label:^iim))  (((!label:^s)  (!label:^k))  (-label:^smartlabel_social)
(-label:^smartlabel_promo))","qa":{}},"vb":0,"M":5,"A":1,"vf":null,"qa":{}},{"ya":
{"U":"^smartlabel_social","name":"PI -
1","color":null,"bgColor":"#4986E7","Vh":2,"fh":false,"Vl":true,"B":null,"A":"((label:^i)
OR  (label:^iim   sp/hybrid/bobrobertson666@gmail.com/imp-  @
      ‣ q
pref/bobrobertson666@gmail.com/ad,K pref/bobrobertson666@gmail.com/uv+_Å
pref/bobrobertson666@gmail.com/ed"K pref/bobrobertson666@gmail.com/cs pref/
bobrobertson666@gmail.com/bx_asnsMpref/bobrobertson666@gmail.com/hns pref/
bobrobertson666@gmail.com/asKpref/bobrobertson666@gmail.com/sd]sp/hybrid/
bobrobertson666@gmail.com/db-rcc?]Å
sp/hybrid/bobrobertson666@gmail.com/imp-ifA] sp/hybrid/bobrobertson666@gmail.com/db-
ptc]sp/hybrid/bobrobertson666@gmail.com/imp-of@bÅI sp/hybrid/bobrobertson666@gmail.com/of-
su-dbn
(!label:^k))
OR (label:^iim)) (((!label:^s) (!label:^k))
:"^io_im","color":null,"bgColor":null,"Vh":null,"fh":fal\=Åsp/hybrid/
bobrobertson666@gmail.com/imp-of  sp/hybrid/bobrobertson666@gmail.com/imp-if©;
0","color":null,"bgColor":"#737373","Vh":1,"fh":false,"Vl":true,"B":null,"A":"(((label:^i)
```

Figure 10-5: Email metadata left in a mail client

Keylogging and the Autocorrection Database

Everyone is familiar with iOS's word autocompletion mechanism, the source of endless entertainment and amusing typos (and of frustration when trying to use expletives). One aspect of this system that's gained some attention in the press is that the autocompletion mechanism acts as an accidental keylogger, recording portions of the text that a user types in what is basically a plaintext file to help with future completions. A forensic attacker could then retrieve that completion database.

This behavior is already disabled for password fields—that is, UITextField objects with setSecureTextEntry:YES set—but many other forms in an application may take sensitive data. As such, developers have to consider the all too common trade-off between user experience and security. For some applications, no amount of unencrypted data stored to disk is acceptable. Other applications handle sensitive data, but they involve so much text entry that disabling autocorrection would be extremely burdensome.

Fields that take smaller amounts of sensitive data, though, are a no-brainer. Consider answers to security questions, for example. For these fields, you'll want to disable autocorrection behavior by setting the autocorrectionType property to UITextAutocorrectionTypeNo on UITextField and UITextView objects. This is also applicable (and a good idea) for UISearchBar objects because having search contents leak to disk is usually undesirable. Check out Listing 10-12 for an example of how you might try to disable this attribute.

```
UITextField *sensitiveTextField = [[UITextField alloc] initWithFrame:CGRectMake(0,
    0, 25, 25)];
[sensitiveTextField setAutocorrectionType:UITextAutocorrectionTypeNo];
```

Listing 10-12: Disabling autocorrection on a UITextField

Of course, note that I say, "You'll want to disable this behavior." You'll *want* to, but you can't. Around iOS 5.1, a bug crept in that causes the on-disk word cache to be updated even if you disable autocorrection, autocapitalization, spellcheck, and so on. There are currently two ways around this, ranging from very silly to utterly ridiculous.

The silly approach (shown in Listing 10-13) is to use a UITextView (note View, rather than Field) and send it the message setSecureTextEntry:YES. The UITextView class doesn't actually implement the UITextInputTraits protocol[9] correctly, so text isn't obscured by circles like it would be in a UITextField configured for password entry. It *does*, however, prevent text from getting written to the disk.

```
-(BOOL)twiddleTextView:(UITextView *)textView {
    [textView setSecureTextEntry:YES];
}
```

Listing 10-13: Setting the SecureTextEntry attribute on a UITextView

The ridiculous method, which works on both UITextView and UITextField objects, is shown in Listing 10-14.

```
-(BOOL)twiddleTextField:(UITextField *)textField {
[textField setSecureTextEntry:YES];
[textField setSecureTextEntry:NO];
}
```

Listing 10-14: Twiddling setSecureTextEntry on a UITextField

Yes, seriously. Just switch keylogging on and then turn it off.

The classes are implemented such that they forget to turn keylogging back on if you simply wiggle it on and off again. Unfortunately, UISearchbar *also* doesn't implement the protocol correctly, so you can't pull this trick on one of the search bars. If preventing data leakage from your search bar is critical, you may want to replace the search bar with an appropriately styled text field instead.

Of course, that bug might be fixed in a future version of the OS, so just be prudent and ensure that the OS your app is running on is a version that you've tested the behavior on before you do this yes/no flipping trick. Listing 10-15 shows how to do this.

```
UITextField *sensitiveTextField = [[UITextField alloc] initWithFrame:CGRectMake(0,
    0, 25, 25)];
[sensitiveTextField setAutocorrectionType: UITextAutocorrectionTypeNo];
```

---

9. *http://developer.apple.com/library/ios/#documentation/uikit/reference/UITextInputTraits_Protocol/Reference/UITextInputTraits.html*

```
if ([[[UIDevice currentDevice] systemVersion] isEqual: @"8.1.4"]) {
    [sensitiveTextField setSecureTextEntry:YES];
    [sensitiveTextField setSecureTextEntry:NO];
}
```

Listing 10-15: Checking iOS version

To help verify that your application isn't leaking anything unexpected, you can also check *<device ID>/data/Library/Keyboard/dynamic-text.dat* on the simulator or on a jailbroken device. (Figure 10-6 shows an example *dynamic-text.dat* file.) This isn't going to catch every corner case of what might be committed to the database, but the file should give you a reasonable idea. Note that the database may not actually get updated until you hit the Home button.

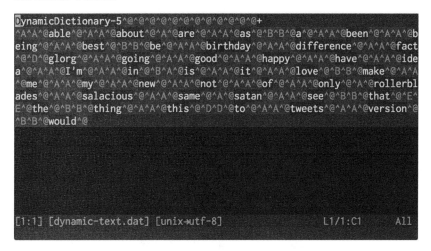

Figure 10-6: Contents of dynamic-text.dat *after using the keyboard to enter message contents. Note that the order of words does not reflect the order in which they were entered.*

In iOS 8 and later, additional information is stored in the Keyboard cache. This data is used to help with the QuickType word prediction system, but it also leaks more information about conversations and people who have communicated with the device owner. In the *<device ID>/data/ Library/Keyboard/en-dynamic.lm* directory,[10] you'll find four additional data files: *dynamic.dat, lexicon.dat, meta.dat,* and *tags.dat.* Check all these files for data entered into your application. Because QuickType adapts based on the recipient, the *tags.dat* file also includes a list of past message recipients so the completion system can use the correct cache for the correct recipient.

───────────────

10. The *en* prefix will be different for different locales, but this is what it is for an English-speaking device.

Misusing User Preferences

As I briefly mentioned in Chapter 3, user preferences often contain sensitive information. But user defaults are actually intended to define, say, what URL an app's API should be at or other nonsensitive preference information.

Preferences are manipulated through the NSUserDefaults API or, less commonly, the CFPreferences API, and many developers clearly must not know what happens to that data on the device. Restrictions on these files are fairly loose, and user preferences can easily be read and manipulated using commonly available tools, such as iExplorer.

Listing 10-16 shows an intentionally terrible usage of NSUserDefaults from the iGoat project.[11]

```
NSUserDefaults *credentials = [NSUserDefaults standardUserDefaults];

[credentials setObject:self.username.text forKey:@"username"];
[credentials setObject:self.password.text forKey:@"password"];
[credentials synchronize];
```

Listing 10-16: The worst possible way to use NSUserDefaults

This is essentially the worst-case scenario for data leakage: the credentials are stored in plaintext in a plist belonging to the app. Many applications in the wild store user credentials this way, and many have been called out for it.

One less common problem with NSUserDefaults is that developers may use it to store information that really shouldn't be under a user's control. For example, some apps hand over the reins for security controls that dictate whether users can download and store files locally or whether they're required to enter a PIN before using the app. To protect users, let the server enforce such decisions as often as possible instead.

When auditing an application, check each use of the NSUserDefaults or CFPreferences APIs to ensure that the data being stored there is appropriate. There should be no secret information or information you don't want a user to change.

Dealing with Sensitive Data in Snapshots

As I also discussed in Chapter 3, iOS snapshots an app's current screen state before sending the app to the background so it can generate an animation when the app is opened again. This results in potentially sensitive information littering the disk, sometimes even if the user doesn't intentionally background the app. For example, if someone happens to answer a call in the middle of entering sensitive information into an application, that screen state will be written to disk and remain there until overwritten with another

---

11. *https://www.owasp.org/index.php/OWASP_iGoat_Project*

snapshot. I've seen many applications willing to record people's SSNs or credit card numbers in this fashion.

Once these snapshots are written to disk, a physical attacker can easily retrieve them with common forensics tools. You can even observe the file being written using the Simulator, as shown in Figure 10-7.

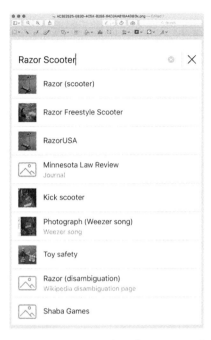

Figure 10-7: A snapshot of a user searching for embarrassing material on Wikipedia, saved to local storage

Just suspend your application and open *UIApplicationAutomaticSnapshot Default-Portrait.png*, which you'll find under your app's *Library/Caches/ Snapshots/com.mycompany.myapp* directory. Unfortunately, applications can't just go and remove snapshots manually. There are, however, a couple of other ways you can prevent this data from leaking.

Screen Sanitization Strategies

First, you can alter the screen state before the screenshot actually occurs. You'll want to implement this in the `applicationDidEnterBackground` delegate method, which is the message that your program receives when the application is going to be suspended, giving you a few seconds to complete any tasks before this occurs.

This delegate is distinct from the `applicationWillResignActive` or `applicationWillTerminate` events. The former is invoked when the application temporarily loses focus (for example, when interrupted by an incoming phone call overlay) and the latter when the application is forcibly killed or has opted out of background operation.[12] For an abbreviated example of the events received over the life cycle of an iOS application, see Figure 10-8.

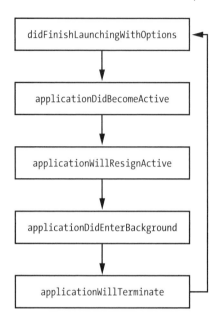

Figure 10-8: The simplified iOS application life cycle. Code for handling these events can be defined in the application delegate.

After these tasks are complete, the snapshot should be taken, and the application should disappear with its little "whoosh" animation. But how can you sanitize your user's screen?

The simplest and most reliable method of obscuring screen contents, and the one that I primarily recommend, is simply placing a splash screen with some logo art on top of all the current views. You can implement this as shown in Listing 10-17.

```
- (void)applicationDidEnterBackground:(UIApplication *)application {

    application = [UIApplication sharedApplication];

    self.splash = [[UIImageView alloc] initWithFrame:[[UIScreen mainScreen]
    bounds]];
```

---

12. For more details on the circumstances under which these events are triggered, visit *http://www.cocoanetics.com/2010/07/understanding-ios-4-backgrounding-and-delegate-messaging/*.

```
[self.splash setImage:[UIImage imageNamed:@"myimage.png"]];
[self.splash setUserInteractionEnabled:NO];
[[application keyWindow] addSubview:splash];
}
```

Listing 10-17: Applying a splash screen

With this code in place, on entering the background, your application should set whatever image you have stored in *myimage.png* as the splash screen. Alternatively, you could set the hidden attribute of the relevant container objects—for example, UITextFields, whose contents might be sensitive. You can use this same approach to hide the entire UIView. This is less visually appealing but easily does the job in a pinch.

A slightly fancier option is to perform some of your own animation,[13] as in Listing 10-18. This just does a fade-out before removing the content from the view.

```
- (void)fadeMe {
    [UIView animateWithDuration:0.2
                     animations:^{view.alpha = 0.0;}
                     completion:^(BOOL finished){[view removeFromSuperview];}
                     ];
}
```

Listing 10-18: Animating a fade to transparency

I even saw one application that took its own screenshot of the current screen state and ran the screenshot through a blur algorithm. It looked pretty, but hitting all the corner cases is tricky, and you'd have to ensure that the blur is destructive enough that an attacker couldn't reverse it.

Regardless of your obfuscation approach, you'll also need to reverse your changes in either the applicationDidBecomeActive or applicationWillEnter Foreground delegate method. For example, to remove the splash image placed over the screen in Listing 10-17, you could add something like Listing 10-19 to the applicationWillEnterForeground method.

```
- (void)applicationWillEnterForeground:(UIApplication *)application {

    [self.splash removeFromSuperview];
    self.splash = nil;
}
```

Listing 10-19: Removing a splash screen

---

13. *http://developer.apple.com/library/ios/#documentation/UIKit/Reference/UIView_Class/UIView/UIView.html#//apple_ref/occ/instp/UIView/alpha*

Before you're done, ensure that your sanitization technique is effective by repeatedly suspending your app in different application states while monitoring your application's *Library/Caches/Snapshots/com.mycompany.myapp* directory. Check that the PNG images saved there have all parts of the window obscured by the splash image.

NOTE   *The* com.mycompany.myapp *directory is re-created on every suspension of the application. If you're watching for the file to be created in that directory from the Terminal, you'll have to reenter the directory using* cd $PWD *or similar for the file to appear.*

Why Do Those Screen Sanitization Strategies Work?

People often misunderstand the fixes I just described because they don't grasp how iOS lays out its views and windows, so I've created a flowchart (Figure 10-9) that shows everything you need to know.

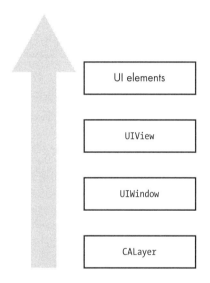

Figure 10-9: The hierarchy of iOS views

Every application that displays contents on the screen is backed by a *layer*, which is CALayer by default. On top of the layer is an instance of the UIWindow class, which manages one or more *views*, instances of the UIView class. UIViews are hierarchical, so a view can have multiple subviews along with buttons, text fields, and so forth.

iOS apps typically have only one UIWindow, but multiple windows are quite possible. By default, windows have a windowLevel property of 0.0, specifying that the window is at the UIWindowLevelNormal level. Other defined levels are UIWindowlevelAlert and UIWindowlevelStatusBar, both of which have level priority over UIWindowLevelNormal, meaning that they'll appear on top of other windows. The most obvious scenario is that of an alert, and in that case, UIAlertView creates a new window on top of all others except the status bar by default.

The window currently receiving user events is referred to as the key window, and it can be referenced via the keyWindow method in UIApplication.

Common Sanitization Mistakes

Developers who don't understand iOS windows and views frequently sanitize screens incorrectly. I've seen several applications that have taken a few development iterations to get it right. One flaw I've seen is to set only the key window's rootViewController to hidden, like so:

```
UIApplication *application;
application = [UIApplication sharedApplication];
[[[[application] keyWindow] rootViewController] view] setHidden:YES];
```

This mistake is understandable because most developers are used to working with UIViews when programming GUIs. While the code will *look* like it works much of the time, it still leaves any subviews of the root visible. An improvement would be to hide the entire key window, like this:

```
UIApplication *application;
application = [UIApplication sharedApplication];
[[[application] keyWindow] setHidden:YES];
```

But hiding the key window isn't a failsafe option either because any UIAlertView windows will appear above other content and become the key window; effectively, you'd end up hiding only the alert.

Because several methods of hiding content are error prone, I almost always recommend that developers use the splash screen approach. There is, however, an even easier, foolproof approach for some use cases: preventing suspension entirely.

Avoiding Snapshots by Preventing Suspension

If your application never really needs to be suspended and resumed (that is, if you want a fresh start with every app launch), then use the Xcode plist editor to add "Application does not run in background" to your plist and set the value to YES, as in Figure 10-10. You can also set UIApplicationExitsOn-Suspend to YES in your *Info.plist* file from your favored text editor.

Adding that item will cause the application to jump to the applicationWill-Terminate event rather than stopping at the applicationDidEnterBackground event, which normally immediately precedes the taking of the screenshot.

Figure 10-10: Adding the "Application does not run in background" item to the plist

Leaks Due to State Preservation

iOS 6 introduced the concept of *state preservation*, which provides a method for maintaining application state between invocations, even if the application is killed in the meantime. When state preservation is triggered, each preservable object's encodeRestorableStateWithCoder delegate method, which contains instructions for how to serialize various UI elements to disk, is called. Then, the decodeRestorableStateWithCoder method is called on relaunch of the application. This system presents a possibility for sensitive information to leak from the user interface to storage on disk since the contents of text fields and other interface data will be put on local storage.

When you are examining a new codebase, you can quickly determine whether any state preservation is happening by grepping the codebase for restorationIdentifier, rather than clicking your way through all the Storyboard UI elements.

If preservation is in use, you should find results like this one in the *.storyboard* files:

```
<viewController restorationIdentifier="viewController2" title="Second" id="3"
    customClass="StatePreservatorSecondViewController" sceneMemberID=
    "viewController">
```

```
<view key="view" contentMode="scaleToFill" id="17">
  <rect key="frame" x="0.0" y="20" width="320" height="499"/>
  <autoresizingMask key="autoresizingMask" widthSizable="YES" heightSizable=
  "YES"/>
  <subviews>
    <textView clipsSubviews="YES" multipleTouchEnabled="YES" contentMode=
    "scaleToFill" translatesAutoresizingMaskIntoConstraints="NO" id="Zl1-t0-jGB">
      <textInputTraits key="textInputTraits" autocapitalizationType=
    "sentences"/>
    </textView>
```

Note that there is a view controller with a `restorationIdentifier` attribute, and this controller contains a subview with a `textView` object. If the application delegate implements the `encodeRestorableStateWithCoder` method, it can specify an `encodeObject` method that preserves the `.text` attribute of the `UITextView` for later restoration. This method can be used to ensure that text typed into the field isn't lost if the application is terminated,[14] as shown in Listing 10-20.

```
-(void)encodeRestorableStateWithCoder:(NSCoder *)coder {
    [super encodeRestorableStateWithCoder:coder];

    [coder encodeObject:_messageBox.text forKey:@"messageBoxContents"];
}
```

Listing 10-20: An example encodeRestorableStateWithCoder *method*

After performing functional testing, you can also examine the application's *Library/Saved Application State/com.company.appname.savedState* directory, where you'll find the descriptively named *data.data* file. This file contains the serialized state of the application for objects that have `restorationIdentifiers` assigned. Examine this file to determine whether any sensitive data from the user interface may have been encoded. You can also do this on the device, if you're performing black-box testing.

Secure State Preservation

If a product needs the UX and convenience of state preservation but needs data to be stored securely while on disk, you can encrypt sensitive object contents before passing them to the `encodeObject` method. I discuss encryption in more detail in Chapter 13), but here's how you'd encrypt this particular sort of data.

---

14. Check out a good example of creating a Storyboard application with state preservation at *http://www.techotopia.com/index.php/An_iOS_6_iPhone_State_Preservation_and_Restoration_Tutorial*.

When the application is installed, generate an encryption key and store it in the Keychain with secItemAdd. Then, in your encodeRestorableStateWith-Coder methods, read the key out of the Keychain and use it as the key for an encryption operation.[15] Take the resulting data and serialize it with the NSCoder's encodeObject method. Finally, in the decodeRestorableStateWithCoder method, perform the same operations in reverse to restore the application's state.

You can use the SecureNSCoder project[16] to help implement that functionality. SecureNSCoder can automatically generate a key for your application, store it in the Keychain, and use it to encode and decode your program state. For the rest of this section, I'll walk you through a sample project that demonstrates how to use this tool in your own programs.

First, include the *SecureArchiveDelegate* and *SimpleKeychainWrapper* files in your project. Then, include *SecureArchiverDelegate.h* in your view controller's *.h* file, as shown in Listing 10-21.

```objc
#import <UIKit/UIKit.h>
#import "SecureArchiverDelegate.h"

@interface ViewController : UIViewController

// Some simple properties, adding one for the delegate
@property (weak, nonatomic) IBOutlet UITextField *textField;
@property (weak, nonatomic) SecureArchiverDelegate *delegate;

@end
```

Listing 10-21: A basic ViewController.h

Next, implement the initWithCoder method, as in Listing 10-22.

```objc
- (id)initWithCoder:(NSKeyedUnarchiver *)coder {
    if (self = [super initWithCoder:coder]) {
        return self;
    }
    return nil;
}
```

Listing 10-22: initWithCoder *in* ViewController.m

---

15. Using CCCrypt or, ideally, RNCryptor: *https://github.com/rnapier/RNCryptor*

16. Available at *https://github.com/iSECPartners/SecureNSCoder*

Then implement the `awakeFromNib` method shown in Listing 10-23.

```
- (void)awakeFromNib {
    self.restorationIdentifier = NSStringFromClass([self class]);
    self.restorationClass = [UIViewController class];
}
```

Listing 10-23: `awakeFromNib` in ViewController.m

Finally, implement the two state preservation methods in Listing 10-24.

```
- (void)encodeRestorableStateWithCoder:(NSKeyedArchiver *)coder {
    // preserve state
    SecureArchiverDelegate *saDelegate = [[SecureArchiverDelegate alloc] init];
    [self setDelegate:saDelegate];
    [coder setDelegate:[self delegate]];
    [coder encodeObject:[[self textField] text] forKey:@"textFieldText"];
    [super encodeRestorableStateWithCoder:coder];
}

- (void)decodeRestorableStateWithCoder:(NSKeyedUnarchiver *)coder {
    // restore the preserved state
    SecureArchiverDelegate *saDelegate = [[SecureArchiverDelegate alloc] init];
    [self setDelegate:saDelegate];
    [coder setDelegate:[self delegate]];
    [[self textField] setText:[coder decodeObjectForKey:@"textFieldText"]];
    [super decodeRestorableStateWithCoder:coder];
}
```

Listing 10-24: Encode and decode methods in ViewController.m

You've seen how data can leak from applications on a device, but what about data that's been backed up to iCloud? Well, if you're dealing with sensitive data, there's really only one technique I can recommend there: avoid storing it on iCloud entirely.

Getting Off iCloud to Avoid Leaks

In recent versions of iOS, much of your application's data can be synced to a user's iCloud account, where it can be shared across devices. By default, only three of your application directories are safe from the clutches of iCloud.

- *AppName.app*
- *Library/Caches*
- */tmp*

If you want any of your other files to remain only on the device, you'll have to take responsibility for them yourself.[17] Set the NSURLIsExcludedFrom-BackupKey attribute on those files, using an NSURL as the path to the file, to prevent the file from backing up to iCloud, as in Listing 10-25.

```
- (BOOL)addSkipBackupAttributeToItemAtURL:(NSURL *)URL {
    NSError *error = nil;

❶   [URL setResourceValue:[NSNumber numberWithBool:YES]
                 forKey:NSURLIsExcludedFromBackupKey
                  error:&error];

    return error == nil;
}
```

Listing 10-25: Setting file attributes to exclude a file from backup

You can set the NSURLIsExcludedFromBackupKey with the setResourceValue NSURL method, shown at ❶.

Closing Thoughts

Data leakage on mobile devices is a broad and ever-changing area that makes up a large percentage of issues found in mobile applications when subjected to security audits. Ideally, some of the things you've examined in this chapter will help you find useful bugs, as well as help you identify changes when newer versions of iOS are released. I'll now move on to cover some basic C and memory corruption attacks, which are usually rarer on iOS but potentially much more dangerous.

---

17. *http://developer.apple.com/library/ios/documentation/iPhone/Conceptual/iPhoneOSProgrammingGuide/iPhoneAppProgrammingGuide.pdf* (page 112)

11

LEGACY ISSUES
AND BAGGAGE FROM C

Objective-C and Cocoa help mitigate many security problems that you might run into with C or C++. Objective-C is, however, still a flavor of C, which fundamentally isn't a "safe" language, and some Cocoa APIs are still vulnerable to the types of data theft or code execution attacks you might expect in C programs. C and C++ can also be intermingled freely with Objective-C. Many iOS applications use large amounts of C and C++ code, whether because developers want to use a familiar library or are trying to keep code as portable as possible between platforms. There are some mitigations in place to prevent code execution attacks, as discussed in Chapter 1, but these can be bypassed by more skilled attackers. As such, it's a good idea to familiarize yourself with these bugs and attacks.

In this chapter, you'll learn about some of the types of attacks to look out for, the places where C bugs creep into Objective-C, and how to fix these issues. The topic of native C code issues is broad, so this chapter is a "greatest hits" of these issues to give you the basic foundation for understanding the theory behind these flaws and the attacks that exploit them.

Format Strings

Format string attacks[1] leverage a misuse of APIs that expect a *format string*, or a string that defines the data types of which the string will be composed. In C, the most commonly used functions that accept format strings are in the `printf` family; there are a number of other functions, such as `syslog`, that accept them as well. In Objective-C, these methods usually have suffixes like `WithFormat` or `AppendingFormat`, though there are several exceptions. Here are examples of all three:

- `[NSString *WithFormat]`
- `[NSString stringByAppendingFormat]`
- `[NSMutableString appendFormat]`
- `[NSAlert alertWithMessageText]`
- `[NSException raise:format:]`
- `NSLog()`

Attackers commonly exploit format string vulnerabilities to do two things: execute arbitrary code and read process memory. These vulnerabilities generally stem from two age-old C format string operators: `%n` and `%x`. The rarely used `%n` operator is meant to store the value of the characters printed so far in an integer on the stack. It can, however, be leveraged to overwrite portions of memory. The `%x` operator is meant to print values as hexadecimal, but when no value is passed in to be printed, it reads values from the stack.

Unfortunately for us bug hunters, Apple has disabled `%n` in Cocoa classes that accept format strings. But the `%n` format string *is* allowed in regular C code, so code execution format string attacks are still possible.[2] The reason that `%n` can result in code execution is because it writes to the stack, and the format string is also stored on the stack. Exploitation varies depending on the specific bug, but the main upshot is that by crafting a format string that contains `%n` and also a memory address to write to, you can get arbitrary integers written to specific parts of memory. In combination with some shell code, this can be exploited similarly to a buffer overflow attack.[3]

The `%x` operator, on the other hand, is alive and well in both Objective-C methods and C functions. If an attacker can pass `%x` to an input that lacks a format string specifier, the input will be interpreted as a format string, and the contents of a stack will be written in hexadecimal where the expected string should appear. If attackers can then view this output, they can collect

---

1. The term *format string attack* was popularized by Tim Newsham's paper of the same name; see *http://www.thenewsh.com/~newsham/format-string-attacks.pdf*.

2. Yes, `%n` works. Xcode might complain about it, but manual builds, such as those performed with the `xcodebuild` command line utility, work fine.

3. You can find more details on exploiting format strings to gain code execution in Scut's paper on the topic; see *https://crypto.stanford.edu/cs155/papers/formatstring-1.2.pdf*.

potentially sensitive information from the process's memory, such as user-names, passwords, or other personal data.

Of course, both of these vulnerabilities rely on a program not control-ling user input properly. Let's take a look at how an attacker might misuse format strings in such a situation and how applications can prevent that from happening.

Preventing Classic C Format String Attacks

The typical example of a format string vulnerability is when a program passes a variable directly to printf, without manually specifying a format string. If this variable's contents are supplied by external input that an attacker can control, then the attacker could execute code on a device or steal data from its memory. You can test some contrived vulnerable code like this in Xcode:

```
char *t;
t = "%x%x%x%x%x%x%x%x";
printf(t);
```

This code simply supplies a string containing a bunch of %x specifiers to the printf function. In a real-world program, these values could come from any number of places, such as a user input field or DNS query result. When the code executes, you should see a string of hexadecimal output written to your console. This output contains the hexadecimal values of variables stored on the stack. If an application has stored a password or encryption key as a value on the stack and parses some attacker-supplied data, an attacker could cause this information to leak to somewhere they can then read. If you change the previous example to contain %n specifiers, the behavior is different. Here's how that would look:

```
char *t;
t = "%n%n%n%n%n";
printf(t);
```

Running this example in Xcode should cause Xcode to drop to lldb with the error EXC_BAD_ACCESS. Whenever you see that message, your program is trying to read or write to some memory it shouldn't. In a carefully crafted attack, you won't see such errors, of course; the code will simply execute.

But you can prevent attackers from hijacking strings pretty easily by controlling user input. In this case, just change that printf to specify its own format string, as follows:

```
char *t;
t = "%n%n%n%n%n";
printf("%s", t);
```

Run this in Xcode, and you should see the literal %n%n%n%n%n written harmlessly to the console. These examples, of course, are plain old C, but knowing how they work will help you explore format string attacks with an Objective-C twist.

Preventing Objective-C Format String Attacks

Similar to plain C, you can pass in any of the printf format operators to a number of different Objective-C APIs. You can test this easily in Xcode by passing a bogus format string to NSLog:

```
NSString *userText = @"%x%x%x%x%x%x%x%x%x%x%x%x%x%x%x%x%x%x%x%x";
NSLog(userText);
```

Much like the previous %x example, this will spit out memory contents in hexadecimal to the console. One format string vulnerability I've come across in real iOS applications is code that passes user-supplied input to a "formatting" function, which does some processing and returns an NSString object, as shown in Listing 11-1.

```
NSString *myStuff = @"Here is my stuff.";
NSString *unformattedStuff = @"Evil things %x%x%x%x%x";
❶ myStuff = [myStuff stringByAppendingFormat:[UtilityClass formatStuff:
        unformattedStuff.text]];
```

Listing 11-1: Totally the wrong way to pass in data to a format string

This example just assumes that the resulting NSString stored in myStuff at ❶ is safe; after all, the contents of *unformattedStuff.text* were "formatted." But unless the formatStuff method has some special way of sanitizing that input file, the resulting string could contain format string specifiers. If that happens, you still have a format string issue, and the resulting string will contain values from the stack.

NSString objects aren't magically safe from format string attacks. The correct way to output an NSString passed to a method requiring a format string is to use the %@ specifier, as shown in Listing 11-2.

```
NSString myStuff = @"Here is my stuff.";
myStuff = [myStuff stringByAppendingFormat:@"%@", [UtilityClass formatStuff:
        unformattedStuff.text]];
```

Listing 11-2: The correct way to use a method expecting a format string

With the %@ specifier in front, no matter how many %x and %n operators *unformattedStuff.text* might contain, myStuff should come out as a harmless string.

The %x and %n specifiers are the most useful ones to attackers, but even in their absence, attackers can still cause undesirable behavior, such as crashes, when trying to read inaccessible memory, even using basic %s specifiers. Now that I've covered how format string attacks work and how to prevent them, I'll show you some other methods of executing malicious code.

Buffer Overflows and the Stack

Buffer overflows have long haunted the world of C, allowing crafted input from untrusted sources to crash programs or execute third-party code within the process of a vulnerable program. While buffer overflows have been known since the 1970s, the first prominent example of their exploitation was the Morris worm, which included a buffer overflow exploit of the UNIX finger daemon.

Buffer overflows start by overwriting portions of memory. The basic memory layout of a process consists of the program code, any data the program needs to run, the stack, and the heap, as shown in Figure 11-1.

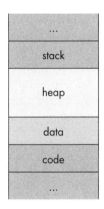

Figure 11-1: Arrangement of process memory

The *code* segment (often referred to as the *text* segment) is where the program's actual executable is loaded into memory. The *data* segment contains the program's global variables and static local variables. The *heap* is where the bulk of nonexecutable program data will reside, in memory dynamically allocated by the program. The *stack* is where local variables are stored, as well as addresses of functions and, importantly, a pointer to the address that contains the next instructions that the program is to execute.

There are two basic types of overflows: those that overwrite portions of a program's stack and those that overwrite portions of the heap. Let's look at a buffer overflow vulnerability now.

A strcpy Buffer Overflow

A classic example of a stack-based buffer overflow is shown in Listing 11-3.

```
#include <string.h>

uid_t check_user(char *provided_uname, char *provided_pw) {
    char password[32];
    char username[32];

    strcpy(password, provided_pw);
    strcpy(username, provided_uname);

    struct *passwd pw = getpwnam(username);

    if (0 != strcmp(crypt(password), pw->pw_passwd))
        return -1;

    return pw->uid;
}
```

Listing 11-3: Code vulnerable to an overflow

Both username and password have been allocated 32 bytes. Under most circumstances, this program should function normally and compare the user-supplied password to the stored password since usernames and passwords tend to be less than 32 characters. However, when either value is supplied with an input that exceeds 32 characters, the additional characters start overwriting the memory adjacent to the variable on the stack, as illustrated in Figure 11-2. This means that an attacker can overwrite the return address of the function, specifying that the next thing to be executed is a blob of malicious code the attacker has placed in the current input or elsewhere in memory.

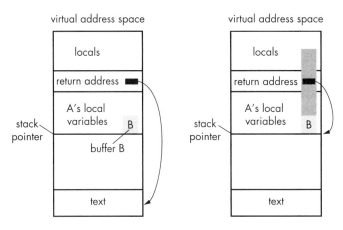

Figure 11-2: Memory layout before and after an overflow

Since this example hardcodes a character limit and doesn't check that the input is within the limit, attacker-controlled input can be longer than the receiving data structure allows. Data will overflow the bounds of that buffer and overwrite portions of memory that could allow for code execution.

Preventing Buffer Overflows

There are a few ways to prevent buffer overflows, and most of them are pretty simple.

Checking Input Size Before Using It

The easiest fix is to sanity check any input before loading it into a data structure. For example, vulnerable programs like the one in Listing 11-3 often defend against buffer overflows by calculating the size of incoming data themselves, rather than trusting an externally supplied size to be the right length. This fix can be as simple as replacing the strcpy functions in Listing 11-3 with if statements like this one:

```
if (strnlen(provided_pw, 32) < strnlen(password, 32))
    strcpy(password, provided_pw);
```

Checking the size of the provided password with sizeof should ensure that any data exceeding the size of the buffer is rejected. Ideally of course, you wouldn't be using statically sized buffers at all—higher level classes like NSString or std::string and their associated methods should take care of these kinds of issues for you.

Using Safer String APIs

Another coding best practice that can protect you from buffer overflows is avoiding "known bad" APIs, such as the strcpy and strcat families. These copy data into a destination buffer without checking whether the destination can actually handle that much data, which is why adding a size check was so important in the previous section. Listing 11-3 showed one bad use of strcpy; here's an even simpler one:

```
void copythings(char *things) {
    char buf[32];
    strcpy(buf, things);
}
```

In this simple and obvious kind of buffer overflow vulnerability, the buf buffer is only 32 bytes long, and the argument things is copied into it. But this code never checks the size of the things buffer before attempting to copy it into buf. If any call this function passes in a buffer is larger than 32 bytes, the result will be a buffer overflow.

The safer way to copy and concatenate strings is to use the strlcpy and strlcat functions,[4] which take the size of the destination buffer as an argument, as follows:

```
void copythings(char *things) {
    char buf[32];
    length = strlcpy(buf, things, sizeof(buf));
}
```

Here, the strlcpy function will copy only 31 bytes of the source string, plus a null terminator. This may result in the string being truncated, but at least it won't overflow the statically sized buffer. The strl family is not available on all platforms but is available on BSD-based systems, including iOS.

In addition to these types of overflows, errors can also be made when performing integer operations, which can lead to a denial of service or code execution.

Integer Overflows and the Heap

Integer overflows result from performing a calculation that gives a value larger than the maximum size of an integer on a platform. As you likely know, there are two types of integers in C (and therefore, in Objective-C): signed and unsigned. Signed integers can be positive or negative, and unsigned integers are always positive. If you attempt to perform a calculation that overflows the value of either type of integers, badness occurs. An unsigned integer will wrap around past the maximum value of an integer, starting over at zero. If the integer is signed, it will start at a negative number, the minimum value of an unsigned integer. Here's an example:

```
NSInteger foo = 9223372036854775807;
NSLog(@"%li", (long)foo);
foo++;
NSLog(@"%li", (long)foo);
```

This starts with a signed integer foo, using the maximum size of a signed integer on iOS. When the number is incremented, the output on the console should wrap around to a negative number, -9223372036854775808.

If you were to use an unsigned integer as shown in the following example, you'd see the integer overflow, and the output on the console would be 0:

```
NSUInteger foo = 18446744073709551615;
NSLog(@"%lu", (unsigned long)foo);
```

---

4. Todd C. Miller, maintainer of sudo, discusses the merits of these functions further at *http://www.sudo.ws/todd/papers/strlcpy.html*.

```
foo++;
NSLog(@"%lu", (unsigned long)foo);
```

While buffer overflows overwrite the stack, integer overflows give attackers access to the heap, and I'll show you how that works next.

A malloc Integer Overflow

An integer overflow most often causes issues when it occurs while calculating the necessary space to pass to a malloc() call, making the space allocated far too small to contain the value to store. When data is loaded into the newly allocated space, the data that won't fit is written beyond the end of the allocated space, into the heap. This puts you in a heap overflow situation: if the attacker provides maliciously crafted data to malloc() and overwrites the right pointer in the heap, code execution can occur.

Integer overflow vulnerabilities tend to take the following form:

```
#define GOAT_NAME_LEN 32

typedef struct Goat {
  int leg_count;     // usually 4
  bool has_goatee;
  char name[GOAT_NAME_LEN];
  struct Goat* parent1;
  struct Goat* parent2;
  size_t kid_count;
  struct Goat** kids;
} Goat;

int ReadInt(int socket) {
  int result;
  read(socket, &result, sizeof(result));
  return result;
}

void ReadGoat(Goat* goat, int socket) {
  read(socket, goat, sizeof(Goat));
}

Goat* ReadGoats(int* count, int socket) {
❶    *count = ReadInt(socket);
❷    Goat* goats = malloc(*count * sizeof(Goat));
❸    for (int i = 0; i < *count; ++i) {
       ReadGoat(&goats[i], socket);
     }
     return goats;
}
```

This code creates an object of type Goat, as well as the ReadGoats function, which accepts a socket and the number of goats to read from that socket. At ❶, the ReadInt function reads the number of goats that will be processed from the socket itself.

If that number is sufficiently large, the malloc() operation at ❷ will result in a size so large that the integer wraps around to negative numbers. With the right value of count, an attacker could make the malloc() attempt to allocate zero bytes, or a very small number. When the loop at ❸ executes, it will read the number of goats from the socket that corresponds to the very large value of count. Because goats is small, this can overflow the allocated memory, allowing data to be written to the heap.

Preventing Integer Overflows

There are several approaches to preventing integer overflows, but the basic idea is to check the values of integers before you operate on them. I suggest adopting the basic structure from Apple's coding guide.[5] Here's one example:

```
if (n > 0 && m > 0 && INT_MAX/n >= m) {
    size_t bytes = n * m;
    foo = malloc(bytes);
}
```

Before calculating the value of bytes, this if statement checks that n and m are greater than 0 and divides one factor by a maximum size to make sure that the result is larger than the other factor. If both conditions are true, then you know that bytes will fit into an integer, and it should be safe to use it to allocate memory.

Closing Thoughts

The list of C coding flaws in this chapter is far from exhaustive, but knowing some of these flaws should help you start spotting C-related issues in iOS applications. There are also many other resources that can help you hone your C security skills. If you're interested in learning more about the intricacies of C and how it can go wrong, I recommend getting a copy of Peter van der Linden's *Expert C Programming: Deep C Secrets* (Prentice Hall, 1994).

Now that I've aired some of the dirty laundry of C, let's head back to Cocoa land and look at modern attacks derived largely from the field of web application security: injection attacks.

---

5. *https://developer.apple.com/library/ios/documentation/Security/Conceptual/SecureCodingGuide/Articles/BufferOverflows.html*

12

INJECTION ATTACKS

In this chapter, I'll discuss types of injection attacks, many of which apply both to iOS client applications and to their remote endpoints or APIs. While a thorough examination of all potential server-side flaws is outside the scope of this book, this chapter will give you an idea of how an iOS app and its complementary endpoint or web app can work together to prevent security flaws.

Injection attacks are standard for web applications, but client-side injection attacks are less common and go largely unnoticed by developers and security engineers. Client-side injection attacks happen when remotely supplied data is parsed by the program running on the device. The most notable examples of this are cross-site scripting, SQL injection, predicate injection, and XML injection.

Client-Side Cross-Site Scripting

Cross-site scripting (XSS) is an issue most often found in web applications, but JavaScript can be injected into the content used by iOS applications, too. One prominent app reported to have an XSS vulnerability was the Skype mobile application. As security researcher Phil Purviance described on his Superevr blog, at the time, the app used a `UIWebView` to render content.[1] The full name of the remote user was not sanitized before display, which allowed an attacker to insert a malicious script into a remote user's application by embedding the script in their username. In this case, the attack could steal sensitive data (the contents of the Address Book) from the device. Such attacks can also be used to, say, insert a fake login page that submits credentials to an attacker-controlled domain.

If your application uses a `UIWebView`, then to avoid XSS vulnerability, be particularly vigilant that you don't take any unsanitized user-supplied data from the server or other external sources and integrate it into the user interface. You can do this most effectively with a two-part approach, using both *input sanitization* and *output encoding*.

Input Sanitization

Input sanitization involves stripping potentially harmful characters from external inputs, using either a *blacklist* or *whitelist* approach.

Blacklisting Bad Input

In a blacklist, you try to list every character that could cause a security problem if accepted as input and give that list to your application. Then, you write your application to either remove unacceptable characters or throw an error when they appear.

Blacklisting is a fragile approach, and it's rarely effective. You need to know every conceivable way data could cause trouble, including every type of character encoding, every JavaScript event handler or SQL special character, and so on. For example, you might simply add < and > to a blacklist in hopes of preventing XSS via `<script>` tags, but you're ignoring attacks that can be accomplished with only double quotes, parentheses, and an equal sign.

In general, if your app or an app you're testing relies on blacklisting characters, investigate whether the blacklist might be masking an underlying flaw. Such filters can be easily bypassed, and an app that relies on this technique probably also lacks effective output encoding, which I'll discuss in "Output Encoding" on page 201.

Whitelisting Allowable Input

In a whitelist approach, you instead explicitly define the characters that are acceptable for a particular user input. Whitelisting is preferable to blacklisting because comprehensively specifying what characters should be allowed is easier than speculating about what might be bad. In a whitelist approach, you might define the characters that a phone number field should allow: 0 through 9 and possibly dashes and parentheses. Not only does this preclude most any malicious input, but it also keeps data clean in your database.

---

1. *https://superevr.com/blog/2011/xss-in-skype-for-ios/*

Finding Balance

It's possible to be misguidedly zealous about input sanitization with either blacklisting or whitelisting. Some programs and websites actually disallow legitimate characters in some inputs (most notably, user passwords). You may have run across an app or site that refuses to accept a password containing special characters (such as !, <, >, ', or ;). This is often an indication that the programmers are handling data on the backend in a remarkably incompetent way.

For example, if an application strips apostrophes or semicolons, the developers may not be using parameterized SQL statements, instead relying on removing "bad" special characters to prevent SQL injection. But this blacklisting of suspected bad characters just reduces user password complexity, and it's unlikely to solve the problem of SQL injection in any comprehensive fashion.

For input sanitization to work correctly, it also needs to happen as close as possible to the point before the data is processed or stored. For example, when an iOS application talks to a remote API, the application can certainly try to strip out harmful characters or restrict input to a certain character range. This is fine, but it *only* results in increased usability for the user. The user can see immediately that their input won't be accepted, rather than waiting until they fill out all the form data and try to submit it.

Your typical users may appreciate that side effect, but there's a problem here: the user controls the device and, ultimately, how your program behaves. If your UI won't allow certain values as input, all an attacker needs to do is route the device's traffic through a proxy, as I described in "Network and Proxy Setup" on page 43. The user can then modify data after it leaves the app but before it reaches the server and add the harmful characters back.

To counter this possibility, never trust a mobile app to supply good data. In a client-server app, always ensure that sanitization happens on the server.

With sane input sanitization in place, you should move on to encoding your output.

Output Encoding

Output encoding, sometimes known as HTML entity encoding, is the process of taking user input and replacing characters with their HTML representations. This process is necessary for any potentially untrusted data that might end up rendered in a WebView. For example, the characters < and > would be translated to < and >, respectively. When data is displayed to the user, those characters should appear in the UI as < and >, but because they've been encoded, the HTML engine doesn't process them as metacharacters, which might be used in a <script> tag.

Output encoding is the last and most potent line of defense before delivering HTML that contains third-party input to a client. Even if you totally neglected to strip potentially harmful metacharacters during input sanitization, as long as you encode your output, you don't have to worry about whether the data you send will be executed by the browser rather than just displayed.

Displaying Untrusted Data

Like input sanitization, output encoding is usually a process you should perform on the server side, not the client. But if you have to display data from domains outside your control that contain untrusted data, you'll want to perform HTML entity encoding before displaying content to the user.

Google Toolbox for Mac includes two category methods of NSString that you could use to encode HTML entities on the client side: gtm_string-ByEscapingForHTML and gtm_stringByEscapingForAsciiHTML.[2] Including Google's category for NSString in your project makes it so you can simply call a method on any NSString object to have it return an encoded representation:

```
NSString *escaped;
escaped = [@"Meet & greet" gtm_stringByEscapingForHTML];
```

After this escaping, escaped should contain the NSString Meet & greet, which should be safe to render within HTML.

Don't Over-Encode

As with input sanitization, be careful not to get carried away with output encoding. Some applications entity-encode received characters before sending them to a server or storing them in a database and then end up reencoding the encoded data. You may have seen the results in mobile apps or web apps.

For example, I once saw an application display a banner inviting me to "Meet & greet." In the underlying HTML source, this data would appear as follows:

```
Meet &amp;amp; greet
```

The original input was already encoded (to &) and would have rendered fine as & in the browser. Encoding it again causes it to show up as & to the user. This doesn't create a security problem, but it can cause your data to become messy and hard to deal with. Just remember that there's a reason the technique is called *output encoding*: it needs to be done just before output.

---

2. You can download Google Toolbox for Mac at *https://code.google.com/p/google-toolbox-for-mac/*.

SQL Injection

Client-side SQL injection results from parsing externally supplied data that injects valid SQL into a badly formed SQL statement. Statements that are constructed dynamically on execution, using unsanitized, externally supplied input, are vulnerable to SQL injection. Malicious input will contain SQL metacharacters and statements that subvert the intent of the original query.

For example, imagine a simple status message is posted to a website by a user. It then gets downloaded and added to a local data store. If the user posting the original content has basic security knowledge and malicious intent, the user could embed SQL into the message, which will be executed when parsed by the SQL engine. This malicious SQL could destroy or modify existing data in the data store.

On iOS, the most commonly used SQL API is SQLite. Listing 12-1 shows an example of an incorrectly formed, dynamically constructed SQL statement for SQLite.

```
NSString *uid = [myHTTPConnection getUID];
NSString *statement = [NSString StringWithFormat:@"SELECT username FROM users where
    uid = '%@'",uid];
const char *sql = [statement UTF8String];
```

Listing 12-1: An unparameterized SQL statement vulnerable to SQL injection

The problem here is that the uid value is being taken from user-supplied input and inserted as is into a SQL statement using a format string. Any SQL in the user-supplied parameter will then become part of that statement when it ultimately gets executed.

To prevent SQL injection, simply use parameterized statements to avoid the dynamic construction of SQL statements in the first place. Instead of constructing the statement dynamically and passing it to the SQL parser, a parameterized statement causes the statement to be evaluated and compiled independently of the parameters. The parameters themselves are supplied to the compiled statement upon execution.

Using parameterized statements, the correct way to structure the query in Listing 12-1 is to use ? as a placeholder character for the supplied parameter, as in Listing 12-2.

```
  static sqlite3_stmt *selectUid = nil;
❶ const char *sql = "SELECT username FROM users where uid = ?";
❷ sqlite3_prepare_v2(db, sql, -1, &selectUid, NULL);
❸ sqlite3_bind_int(selectUid, 1, uid);
  int status = sqlite3_step(selectUid);
```

Listing 12-2: A properly parameterized SQL statement

The SQL statement is constructed with the ? placeholder at ❶. The code then compiles the SQL statement with sqlite3_prepare_v2 at ❷ and lastly binds the user-supplied uid using sqlite3_bind_int at ❸. Since the SQL statement has already been constructed, no additional SQL provided in the uid parameter will be added to the SQL itself; it's simply passed in by value.

In addition to preventing SQL injection, using parameterized, prepared statements will improve application performance under most circumstances. You should use them for all SQL statements, even if a statement isn't taking input from untrusted sources.

Predicate Injection

Predicates let you perform logical comparisons between data using a basic query language not dissimilar to SQL. In a basic NSPredicate, values are compared or filtered using format strings.

```
❶ NSMutableArray *fruit = [NSMutableArray arrayWithObjects:@"Grape", @"Peach",
       @"orange", @"grapefruit", nil];
❷ NSPredicate *pred = [NSPredicate predicateWithFormat:@"SELF CONTAINS[c] 'Grape'"];
❸ NSArray *grapethings = [fruit filteredArrayUsingPredicate:pred];
   NSLog(@"%@", grapethings);
```

At ❶, an array of various types of fruit is created; this array will be the data source to evaluate against an expression. When creating a predicate at ❷, a query is created that checks whether the string "Grape" is contained in the item the predicate is being compared to. (The [c] makes this comparison case insensitive.) When a new array is instantiated at ❸ to contain the results of this comparison, the filteredArrayUsingPredicate method of the fruit array is used to pass in the predicate. The resulting grapethings array should now contain both "Grape" and "grapefruit".

So far, so good! But a few things can go wrong when you build a predicate query using externally supplied data. First, consider the case where a predicate is built using SQL's LIKE operator, as follows.

```
NSPredicate *pred;
pred = [NSPredicate predicateWithFormat:@"pin LIKE %@", [self.pin text]];
```

This example evaluates a PIN, perhaps a secondary form of authentication for my application. But the LIKE operator performs the evaluation, which means a simple entry of the wildcard character (*) from a user will cause the predicate to evaluate to true, effectively bypassing PIN protection.

This result may seem obvious to those familiar with SQL injection (since SQL also has a `LIKE` operator), but consider the more subtle case where you're examining code that uses the predicate `MATCHES` operator, as shown here:

```
NSPredicate *pred;
pred = [NSPredicate predicateWithFormat:@"pin MATCHES %@", [self.pin text]];
```

This code has the same issue as the `LIKE` example, but rather than just accepting wildcards, `MATCHES` expects a regular expression. Therefore, using `.*` as your PIN will be enough to bypass validation.

To prevent predicate injection attacks, examine all uses of `NSPredicate` in your code and make sure that the operators being used make sense for the application. It's also probably a good idea to limit the characters that are allowed in user-supplied data that gets passed to a predicate to ensure that characters like wildcards don't get plugged in. Or, simply don't use a predicate for security-sensitive operations.

XML Injection

XML injection occurs when malicious XML is parsed by an XML parser instance. Typically, this type of attack is used to force an application to load external resources over the network or consume system resources. In the iOS world, the most commonly used XML parser is the Foundation `NSXMLParser` class.

Injection Through XML External Entities

One basic function of an XML parser is to handle XML entities. You can basically think of these as shortcuts or euphemisms. For example, say you have a simple string like this one:

```
<!ENTITY myEntity "This is some text that I don't want to have to spell out
    repeatedly">
```

You could then reference the entity in other parts of an XML document, and the parser would insert the contents of the entity at that placeholder. To reference your defined entity, simply use this syntax:

```
<explanation>&myEntity;</explanation>
```

NSXMLParser instances have several configurable parameters that can be set after instantiation. If shouldResolveExternalEntities is set to YES on an NSXMLParser instance, the parser will honor *Document Type Definitions (DTDs)*, which can define entities fetched from external URLs. (That's why these are called *external* entities.) When a defined entity is encountered later in the parsed XML, the URL will be requested, and the results of the query will be used to populate the XML, as in this example:

```
NSURL *testURL = [NSURL URLWithString:@"http://api.nostarch.com"];
NSXMLParser *testParser = [[NSXMLParser alloc] initWithContentsOfURL:testURL];
[testParser setShouldResolveExternalEntities:YES];
```

Here, an XML parser is instantiated that reads data from an NSURL passed to the initWithContentsOfURL argument. But if the remote server decides to return huge amounts of data, or to simply hang, the client application may crash or hang in response.

Remember, however, that an external entity can also refer to a local file, meaning the file's contents could be included in your parsed XML. If that XML is stored and then later delivered to the server or another third party, the contents of the file will be disclosed along with the rest of the XML. To avoid such scenarios, ensure that any URL or filename passed to the XML parser is thoroughly sanitized, ideally by a using whitelisting approach, as I discussed in relation to cross-site scripting in "Whitelisting Allowable Input" on page 12.

Note that in iOS 7.0 and 7.1 the default behavior of the XML parser is to resolve external entities (the opposite of the parser's intended behavior), and using setShouldResolveExternalEntities:NO doesn't actually work.[3] Unfortunately, there is no workaround to secure the XML parser for older versions of iOS, short of using an alternative XML parser. The issue was resolved in iOS 8.

NOTE *Contrary to what some have claimed,* NSXMLParser *is not vulnerable to recursive entity attacks, a type of denial of service otherwise known as the* billion laughs *attack. Vulnerable parsers will resolve recursive entities (entities that reference other entities) and chew up tons of system resources. However, if recursive entity declarations are given to* NSXMLParser*, an* NSXMLParserEntityRefLoopError *is thrown.*

Misuse of official external entities isn't the only element of XML injection to watch for in iOS code, however. Some apps incorporate third-party XML libraries, which bring their own set of problems.

---

3. *http://support.apple.com/kb/HT6441*

Issues with Alternative XML Libraries

You may encounter alternative XML libraries in various iOS projects, generally chosen for their improved performance characteristics over NSXMLParser and their support for features such as XPath. (Ray Wenderlich offers a good tutorial on choosing an XML parser on his blog.[4]) When examining code that uses an alternate XML library, first ensure that external entity expansion is disabled using that library's standard methods. Then, confirm that any XPath queries that integrate externally supplied input sanitize the input first, as you would when preventing cross-site scripting. XPath queries should also be parameterized in a manner similar to that of SQL queries (see "SQL Injection" on page 203), but the methods for doing this may vary depending on which third-party libraries are involved.

Closing Thoughts

Ultimately, handling most of the attacks in this chapter comes down to treating all external input as hostile: remove potentially malicious content and encode or prepare it, if possible, to prevent code execution. It's a good idea to be specific about the content that is allowed for each parameter fetched from the UI or from a remote user-manipulated source and enforce this in your program.

Now I'll turn away from shielding against malicious data and toward protecting good data with appropriate cryptography.

---

4. *http://www.raywenderlich.com/553/how-to-chose-the-best-xml-parser-for-your-iphone-project*

PART IV

KEEPING DATA SAFE

13

ENCRYPTION AND AUTHENTICATION

While Apple's cryptographic APIs are fairly robust, many developers don't know how to use them effectively. There are two major built-in encryption components that you have control over: the Keychain and the Data Protection API. These components share some of the same encryption keys and have similar protection attributes, and I'll cover them in this chapter. I'll also provide a look at lower-level crypto primitives and the (limited) circumstances in which you would want to use them.

Using the Keychain

The Keychain is meant to be used when you have small snippets of sensitive data to store, including passwords, personal data, and so on. The Keychain itself is encrypted using the Device Key, combined with a user passcode if available. The Keychain's API consists of four main operations: `SecItemAdd`, `SecItemUpdate`, `SecItemCopyMatching`, and `SecItemDelete`. These operations add items to the Keychain, update existing items, retrieve items, and delete them from the Keychain, respectively.

That said, I *really* wish I'd never see the GenericKeychain[1] sample code again. Everyone seems to base their Keychain code on it (which is reasonable), but this code predates any of the modern Keychain protections that actually prevent secret data from being stolen off your device by a physical attacker. In this section, you'll learn about those protections and how to take advantage of them.

The Keychain in User Backups

When users perform full backups of their devices, they have two security-related options: Unencrypted and Encrypted. *Unencrypted* backups can be restored only to the same device they were received from. *Encrypted* backups let the user select a passphrase to encrypt their backup data with. This allows the backup to be restored to any device (except for items marked with ThisDeviceOnly) and backs up the full contents of the Keychain as well. If you don't want your Keychain item to be stored in backups, you can use the Keychain's data protection attributes.

Keychain Protection Attributes

Keychain protection attributes specify when Keychain data is allowed to be stored in memory and requested by the OS or an application. When adding items such as passwords or personal data to the Keychain, it's important to specify a protection attribute because this explicitly states when the data should be available. Not specifying a protection attribute should be considered a bug.

Specify attributes when first storing an item in the Keychain by using the SecItemAdd method. You'll need to pass in one of a predefined set of values (see Table 13-1) for kSecAttrAccessible.

Three main types of access can be specified via this attribute:

Always accessible   The key is always available, regardless of whether the phone is locked.

Accessible when unlocked   The key is accessible when the device is unlocked; otherwise, attempts to access it will fail.

Accessible after first unlocked   The key is accessible after the device has booted and been unlocked for the first time.

For each of the three main types of Keychain protection, there is an additional counterpart suffixed with ThisDeviceOnly. This means that the Keychain item will not be backed up to iCloud, will be backed up to iTunes only if using encrypted backups, and cannot be restored onto another device.

---

1. *http://developer.apple.com/library/ios/#samplecode/GenericKeychain/Introduction/Intro.html*

Table 13-1: Keychain Protection Attributes and Their Associated Meanings

Keychain protection attribute	Meaning
`kSecAttrAccessibleAfterFirstUnlock`	The key is inaccessible after boot, until the user enters a passcode for the first time.
`kSecAttrAccessibleAlways`	The key is always accessible, as long as the device is booted. Note that this is deprecated in iOS 9 because it has no real advantage over `kSecAttrAccessibleAfterFirstUnlock`.
`kSecAttrAccessibleAlwaysThisDeviceOnly`	The key is always accessible, but it cannot be ported to other iOS devices.
`kSecAttrAccessibleAfterFirstUnlockThisDeviceOnly`	This is the same as the previous key, but this key remains on only this device.
`kSecAttrAccessibleWhenUnlocked`	Whenever the device is unlocked (that is, after the user has entered a passcode), the key is accessible.
`kSecAttrAccessibleWhenUnlockedThisDeviceOnly`	This is the same as the previous key, but this key remains only on this device (except for full, encrypted backups).
`kSecAttrAccessibleWhenPasscodeSetThisDeviceOnly`	This is the same as the previous key, but this key will be available only to users who have a passcode set and will be removed from the device if that passcode is unset. It will not be included in any backups.

When Keychain protections were first introduced, the default value was kSecAttrAccessibleAlways, creating an obvious security problem. *Accessible* in this case should be taken to mean "available to a physical attacker": if someone steals your device, they'll be able to read the contents of the Keychain. Generally, this is done by performing a temporary jailbreak and extracting the keys; using kSecAttrAccessibleAfterFirstUnlock instead will usually prevent this since a reboot is often required to perform the jailbreak. However, a code execution attack (such as someone exploiting a bug in a Wi-Fi driver) would give access to a device while it's still running. In this case, kSecAttrAccessibleWhenUnlocked would be needed to prevent compromise of the keys, meaning that the attacker would need to determine the user's passcode to extract secrets.

Unfortunately, brute-forcing a four-digit PIN on iOS is ridiculously fast. Not only can this be done with a temporary jailbreak,[2] but my colleagues have successfully built cute robots to physically brute-force PINs in less than a day.[3]

Currently, the default attribute is kSecAttrAccessibleWhenUnlocked, which is a reasonably restrictive default. However, Apple's public documentation disagrees about what the default attribute is supposed to be, so just in case, you should set this attribute explicitly on all Keychain items. For your own code, consider using kSecAttrAccessibleWhenUnlockedThisDeviceOnly if appropriate; when examining third-party source code, ensure that restrictive protection attributes are used.

In iOS 8, the kSecAttrAccessibleWhenPasscodeSetThisDeviceOnly protection attribute was added. Developers have long requested an API that requires a user to have a passcode set. This new attribute doesn't directly accomplish that, but developers can use it to make decisions based on whether a passcode is set. When you attempt to add an item to the Keychain specifying the kSecAttrAccessibleWhenPasscodeSetThisDeviceOnly attribute, it will fail if the user does not have a passcode set. You can use this failure as a point at which to make a decision about whether to fall back to another Keychain protection attribute, alert the user, or just store less sensitive data locally.

If the user does have a passcode set, the addition will be successful; however, if the user ever decides to disable the passcode, the Class Keys used to decrypt the item will be discarded, preventing the item from being decrypted by the application.

Basic Keychain Usage

There are several classes of Keychain items, as listed in Table 13-2. Unless you're dealing with certificates, kSecClassGenericPassword can generally be used for most sensitive data, so let's look at some useful methods on that class.

Table 13-2: Keychain Item Classes

Item class	Meaning
kSecClassGenericPassword	A plain-old password
kSecClassInternetPassword	A password specifically used for an Internet service
kSecClassCertificate	A cryptographic certificate
kSecClassKey	A cryptographic key
kSecClassIdentity	A key pair, comprising a public certificate and private key

---

2. *https://www.trailofbits.com/resources/ios4_security_evaluation_slides.pdf*

3. *https://github.com/iSECPartners/R2B2*

Listing 13-1 shows an example of how to use the Keychain to add a basic password item, using SecItemAdd. It sets up a dictionary to hold a Keychain *query*, which contains the appropriate key-value pairs to identify the password, sets a password policy, and specifies the password itself.

```
NSMutableDictionary *dict = [NSMutableDictionary dictionary];
NSData *passwordData = [@"mypassword" dataUsingEncoding:NSUTF8StringEncoding];

[dict setObject:(__bridge id)kSecClassGenericPassword forKey:(__bridge id)
    kSecClass];
[dict setObject:@"Conglomco login" forKey:(__bridge id)kSecAttrLabel];
[dict setObject:@"This is your password for the Conglomco service." forKey:
    (__bridge id)kSecAttrDescription];
[dict setObject:@"dthiel" forKey:(__bridge id)kSecAttrAccount];
[dict setObject:@"com.isecpartners.SampleKeychain" forKey:(__bridge id)
    kSecAttrService];
[dict setObject:passwordData forKey:(__bridge id)kSecValueData];
[dict setObject:(__bridge id)kSecAttrAccessibleWhenUnlocked forKey:(__bridge id)
    kSecAttrAccessible];

OSStatus error = SecItemAdd((__bridge CFDictionaryRef)dict, NULL);
if (error == errSecSuccess) {
    NSLog(@"Yay");
}
```

Listing 13-1: Adding an item to the Keychain

Here, the kSecClassGenericPassword class is set for the Keychain item, along with a user-readable label, a long description, the account (username), and an identifier for the service (to prevent duplicates). The code also sets the password and an accessibility attribute.

SecItemUpdate works similarly. Listing 13-2 shows SecItemUpdate in action with an example that updates the user's password, which is stored in kSecValueData.

```
NSString *newPassword = @"";
NSMutableDictionary *dict = [NSMutableDictionary dictionary];

[dict setObject:(__bridge id)kSecClassGenericPassword forKey:(__bridge id)
    kSecClass];
[dict setObject:@"dthiel" forKey:(__bridge id)kSecAttrAccount];
[dict setObject:@"com.isecpartners.SampleKeychain" forKey:(__bridge id)
    kSecAttrService];

NSDictionary *updatedAttribute = [NSDictionary dictionaryWithObject:[newPassword
    dataUsingEncoding:NSUTF8StringEncoding] forKey:(__bridge id)kSecValueData];
```

```
OSStatus error = SecItemUpdate((__bridge CFDictionaryRef)dict, (__bridge
    CFDictionaryRef)updatedAttribute);
```

Listing 13-2: Updating a Keychain item with SecItemUpdate

When updating a Keychain item with SecItemUpdate, you have to set two dictionaries. One should specify the basic Keychain identification information (at least the class, account, and service information), and the other should contain the attribute to update.

SecItemCopyMatching can be used to query the Keychain to find one or more entries matching a given set of criteria. Typically, you'd construct a search dictionary using the class, account, and service attributes you use when creating or updating a Keychain item. Then, you'd instantiate an NSDictionary that will hold the search results and perform the actual SecItem-CopyMatching call, passing in the search dictionary and a reference to the result dictionary. An example can be found in Listing 13-3.

```
[dict setObject:(__bridge id)kSecClassGenericPassword forKey:(__bridge id)
    kSecClass];
[dict setObject:@"dthiel" forKey:(__bridge id)kSecAttrAccount];
[dict setObject:@"com.isecpartners.SampleKeychain" forKey:(__bridge id)
    kSecAttrService];

[dict setObject:(id)kCFBooleanTrue forKey:(__bridge id)kSecReturnAttributes];

NSDictionary *result = nil;
OSStatus error = SecItemCopyMatching((__bridge CFDictionaryRef)dict, (void *)&
    result);
NSLog(@"Yay %@", result);
```

Listing 13-3: Querying the Keychain using SecItemCopyMatching

With the Keychain data in the result dictionary, you can then use this information to perform your security-sensitive tasks such as authenticating to a remote service or decrypting data. Note that if you construct a query based on attributes that don't include the account and service (which uniquely identify Keychain items), you may get a return dictionary that contains more than one Keychain item. This dictionary can be limited with kSecMatchLimit (that is, by setting it to a value of 1), but this could lead to unpredictable behavior if you're trying to search for a single piece of data like a password.

You can probably guess at this point what a SecItemDelete call will look like—see the example in Listing 13-4 for the actual code.

```
NSMutableDictionary *searchDictionary = [NSMutableDictionary dictionary];

[searchDictionary setObject:(__bridge id)kSecClassGenericPassword forKey:
    (__bridge id)kSecClass];
[searchDictionary setObject:@"dthiel" forKey:(__bridge id)kSecAttrAccount];
[searchDictionary setObject:@"com.isecpartners.SampleKeychain" forKey:(__bridge id)
    kSecAttrService];

OSStatus error = SecItemDelete((__bridge CFDictionaryRef)searchDictionary);
```

Listing 13-4: Deleting a Keychain item using SecItemDelete

Note that if you don't uniquely identify your Keychain item, all matching items that your application has access to will be deleted.

Keychain Wrappers

When working with the Keychain, you'll probably end up writing a number of wrapper functions to make it more convenient since most applications use only a subset of the Keychain API's functionality. There are actually a number of prewritten Keychain wrappers available from third parties; I tend to prefer Lockbox[4] for its simplicity and functionality. Lockbox provides a set of class methods for storing strings, dates, arrays, and sets. You can see the procedure for storing a secret string in Listing 13-5.

```
#import "Lockbox.h"

NSString *keyname = @"KeyForMyApp";
NSString *secret = @"secretstring";

[Lockbox setString:secret
           forKey:keyname
    accessibility:kSecAttrAccessibleWhenUnlocked];
```

Listing 13-5: Setting a Keychain item with Lockbox

The key name will be prefixed with your application's bundle ID automatically, and this value will be used as for both the account and service keys. Retrieving data from the Keychain works as shown in Listing 13-6.

```
NSString *result = [Lockbox stringForKey:secret];
```

Listing 13-6: Retrieving a string from the Keychain using Lockbox

---

4. *https://github.com/granoff/Lockbox*

Whichever wrapper you choose or write, ensure that it has the ability to set kSecAttrAccessible attributes because much available sample code neglects this feature.

Shared Keychains

iOS has the capability to share Keychain data among multiple applications from the same developer by using Keychain access groups. For example, if you have a "buyer" app and a "seller" app for an online marketplace, you can let your users share the same username and password between the two applications. Unfortunately, this mechanism is widely misunderstood, which has led people to do horrifying things such as using named pasteboards to share items that should be specific to the Keychain.

NOTE *To use Keychain access groups, your applications* must *share the same bundle seed ID. This can be specified only upon creation of a new App ID.*[5]

For your application to take advantage of access groups, you'll need to create an Entitlements property list (see Figure 13-1) containing an array called keychain-access-groups, with a String entry for each shared Keychain item.

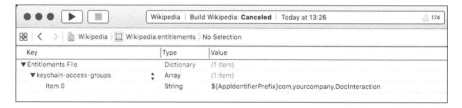

Figure 13-1: Define a Keychain access group consisting of your bundle seed ID and your company prefix, followed by a common name for the Keychain item.

The Keychain item will consist of the bundle seed ID, followed by your reverse-DNS notation developer identifier and a symbolic name that both applications can refer to the entitlement with (see Listing 13-7).

```
[dict setObject:@"DTHIELISEC.securitySuite" forKey:(id)kSecAttrAccessGroup];
```

Listing 13-7: Setting the access group of a Keychain item

---

5. *http://useyourloaf.com/blog/2010/4/3/keychain-group-access.html*

Here, `DTHIELISEC` is the bundle seed ID. Your bundle ID will also be a 10-character alphanumeric string. You'll need to pass in the value of your new entitlement as the value of the `kSecAttrAccessGroup` key when creating a Keychain item via the `SecItemAdd` function. Note that you can have only one Keychain access group on a Keychain item.

NOTE   *Technically, if you create a Keychain access group and don't specify it when creating a Keychain item, the first string in the `keychain-access-groups` array will be used as the default entitlement. So if you're using only one access group, you don't have to specify the group when doing a `SecItemAdd`—but you should anyway.*

iCloud Synchronization

iOS 7 introduced a mechanism to allow Keychain items to be synchronized with iCloud, letting users share Keychain items across multiple devices. By default, this is not enabled on application-created Keychain items, but it can be enabled by setting `kSecAttrSynchronizable` to true.

```
[query setObject:(id)kCFBooleanTrue forKey:(id)kSecAttrSynchronizable];
```

Because this item is now potentially synchronized between multiple Keychains, updates to the item (including deletion) will propagate to all other locations as well. Ensure that your application can handle having Keychain items removed or changed by the system. Also note that you can't specify an incompatible `kSecAttrAccessible` attribute when using this option. For instance, specifying `kSecAttrAccessibleWhenUnlockedThisDeviceOnly` doesn't work because `ThisDeviceOnly` specifies that the item should never be backed up, either to iCloud, to a laptop or desktop, or to any other synchronization provider.

The Data Protection API

As an extra layer of protection, Apple introduced the Data Protection API (not to be confused with Microsoft's Data Protection API), which allows developers to specify when file decryption keys are available. This lets you control access to the file itself, similar to the `kSecAttrAccessible` attribute of Keychain items. The Data Protection API uses the user's passcode in conjunction with a Class Key to encrypt keys specific to each protected file and discards the Class Key in memory when those files should not be accessible (that is, when the device is locked). When a PIN is enabled, the passcode settings screen will indicate that Data Protection is enabled, as in Figure 13-2.

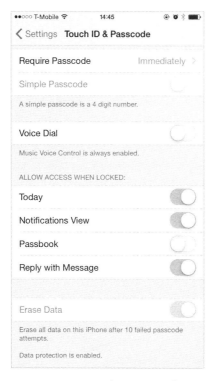

Figure 13-2: Passcode settings with Data Protection enabled

Protection Levels

There are several levels of protection that a developer can request with the Data Protection API, which are roughly analogous to the kSecAttrAccessible attributes one sets on Keychain items. Let's explore those now.

The CompleteUntilFirstUserAuthentication Protection Level

CompleteUntilFirstUserAuthentication is the default file protection attribute for iOS 5 and later. It will be applied to all applicable files unless another attribute has been explicitly specified. It's functionally similar to File-ProtectionComplete, except the file is always available after the user first unlocks the device after a reboot. This doesn't offer a ton of protection if someone gains remote code execution on your running device or if there's a Sandbox bypass, but it does protect you from some attacks that require a reboot.

The Complete Protection Level

Complete is the safest file protection class available, if you can get away with using it. Complete protection ensures that after a short delay, locking the device discards the Class Key from memory and renders file contents unreadable.

This protection level is expressed with the NSFileProtectionComplete attribute of NSFileManager and the NSDataWritingFileProtectionComplete flag for NSData objects. For NSData objects, you can start by setting the NSDataWritingFileProtectionComplete flag, as shown in Listing 13-8.

```
NSData *data = [request responseData];

if (data) {
    NSError *error = nil;
    NSString *downloadFilePath = [NSString stringWithFormat:@"%@mydoc.pdf",
    NSTemporaryDirectory()];
    [data writeToFile:downloadFilePath options:NSDataWritingFileProtectionComplete
        error:&error];
```

Listing 13-8: Setting the NSDataWritingFileProtectionComplete flag on an NSData object

Once you've set NSDataWritingFileProtectionComplete on your NSData object, you can use NNSFileManager to set the NSFileProtectionComplete flag.

```
NSArray *searchPaths = NSSearchPathForDirectoriesInDomains(NSDocumentDirectory,
    NSUserDomainMask, YES);
NSString *applicationDocumentsDirectory = [searchPaths lastObject];
NSString *filePath = [applicationDocumentsDirectory stringByAppendingPathComponent:
    @"mySensitivedata.txt"];

NSError *error = nil;
NSDictionary *attr =
    [NSDictionary dictionaryWithObject:NSFileProtectionComplete
                                forKey:NSFileProtectionKey];

[[NSFileManager defaultManager] setAttributes:attr
                            ofItemAtPath:filePath
                                   error:&error];
```

Listing 13-9: Setting the NSFileProtectionComplete flag using NSFileManager

You can also add file protection attributes on SQLite databases that you create, using the weirdly long SQLITE_OPEN_READWRITE_SQLITE_OPEN_FILEPROTECTION_COMPLETEUNLESSOPEN argument, as shown in Listing 13-10.

```
NSString *databasePath = [documentsDirectory stringByAppendingPathComponent:@"
    MyNewDB.sqlite"];

sqlite3_open_v2([databasePath UTF8String], &handle, SQLITE_OPEN_CREATE|
    SQLITE_OPEN_READWRITE_SQLITE_OPEN_FILEPROTECTION_COMPLETEUNLESSOPEN,NULL);
```

Listing 13-10: Setting protection attributes on SQLite databases

Think about how your app works before trying to use complete protection. If you have a process that will need continuous read/write access to a file, this protection mode will not be appropriate.

The CompleteUnlessOpen Protection Level

The CompleteUnlessOpen protection level is slightly more complicated. You'll set it with the NSFileProtectionCompleteUnlessOpen flag when using NSFileManager and set it with NSDataWritingFileProtectionCompleteUnlessOpen when manipulating NSData stores. It is not, as its name might suggest, a mechanism that disables file protection if a file is currently held open by an application. CompleteUnlessOpen actually ensures that open files can still be written to after the device is locked and allows new files to be written to disk. Any existing files with this class cannot be opened when the device is locked unless they were already open beforehand.

The way this works is by generating a key pair and using it to calculate and derive a shared secret, which wraps the file key. Figure 13-3 illustrates this process.

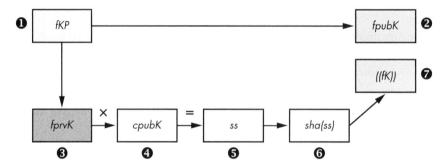

Figure 13-3: Key generation and wrapping. Note that the file private key ❸ is transient and is discarded after the wrapped file key is stored in the file metadata.

Let's walk through this file protection process step by step:

1. As with all files, a file key ❶ is generated to encrypt the file's contents.

2. An additional key pair is generated[6] to produce the file public key ❷ and the file private key ❸.

3. The file private key ❸ and the Protected Unless Open class public key ❹ are used to calculate a shared secret ❺.

4. An SHA-1 hash of this secret ❻ is used to encrypt the file key.

5. This encrypted file key ❼ is stored in the file's metadata, along with the file public key.

6. The system throws away the file private key.

7. Upon closing the file, the unencrypted file key is erased from memory.

---

6. iOS generates the file public and file private keys using D. J.Bernstein's Curve25519, an Elliptic Curve Diffie-Hellman algorithm (*http://cr.yp.to/ecdh.html*).

8. When the file needs to be opened again, the Protected Unless Open class private key and the file public key are used to calculate the shared secret.

9. The SHA-1 hash of this key is then used to decrypt the file key.

The upshot of this process is that you can still write data while the device is locked, without having to worry that an attacker will be able to read that data.

The DataProtectionClass Entitlement

If your application is not going to need to read or write any files while in the background or when the device is locked, you can add an entitlement to the project with a value of NSFileProtectionComplete. This will ensure that all protectable data files written to disk will be accessible only when the device is unlocked, which is the equivalent of setting kSecAttrAccessibleWhenUnlocked on every applicable file.

NOTE   *While this will affect files managed with NSFileManager, NSData, SQLite, and Core Data files, other types of files (for example, plists, caches, and so on) will not be protected.*

In Xcode 5 and later, the Data Protection entitlement is enabled by default on new projects; however, old projects will not be automatically migrated. Enabling entitlement itself is fairly simple—just flip the switch as shown in Figure 13-4.

Figure 13-4: Enable the Data Protection entitlement in Xcode 5.

Note that applications that were installed before iOS 7 do not automatically have Data Protection enabled. They either need to be updated or must have specifically requested a Data Protection attribute in the past.

Checking for Protected Data Availability

For applications that do all their work in the foreground, Data Protection should work transparently. For applications that need to work in the background while the device is locked, your application will need to determine the availability of protected data before using it. This can be done via three different mechanisms.

Implementing Delegate Methods

To be notified and take particular actions when data's availability changes, your application should implement the `applicationProtectedDataWillBecome-Unavailable` and `applicationProtectedDataDidBecomeAvailable` delegate methods, as in Listing 13-11.

```
- (void)applicationProtectedDataWillBecomeUnavailable:
    (UIApplication *)application {

    [self [theBodies hide]];
}

- (void)applicationProtectedDataDidBecomeAvailable:
    (UIApplication *)application {

    [self [theBodies exhume]];
}
```

Listing 13-11: Delegate methods for detecting Data Protection availability changes

Use these delegate methods to ensure that tasks requiring protected data files clean up gracefully and to notify you when the files will be active again.

Using the Notification Center

The `NSNotificationCenter` API essentially allows for an in-app broadcast mechanism, where one part of the app can listen for an event notification that can be called from other places in the code. To use the Notification Center to detect these state changes, you can register for the `UIApplicationProtected-DataWillBecomeUnavailable` and `UIApplicationProtectedDataDidBecomeAvailable` notifications, as shown in Listing 13-12.

```
- (BOOL)application:(UIApplication*)application didFinishLaunchingWithOptions:
    (NSDictionary*)launchOptions {
```

```
❶  NSNotificationCenter *nc = [NSNotificationCenter defaultCenter];

❷  [nc addObserver:self
         selector:@selector(dataGoingAway:)
            name:UIApplicationProtectedDataWillBecomeUnavailable
          object:nil];
   }

❸  - (void)dataGoingAway {

       [self [theBodies hide]];

   }
```

Listing 13-12: Using the Notification Center to detect data availability changes

At ❶, an instance of the default Notification Center is instantiated and then an observer ❷ is added that specifies the selector to call when the event specified by name: occurs. Then you can simply implement that selector as part of the same class ❸ and put any logic that you want to perform upon receipt of the event there.

Detecting Data Protection Using UIApplication

You can also easily detect whether Data Protection is engaged at any given time, as shown in Listing 13-13.

```
if ([[UIApplication sharedApplication] isProtectedDataAvailable]) {

    [self [theBodies hide]];
}
```

Listing 13-13: Using the protectedDataAvailable *property*

Just check the Boolean result of the isProtectedDataAvailable instance method of UIApplication.

Encryption with CommonCrypto

First things first: you are (probably) not a cryptographer.[7] I'm not a cryptographer. It's easy to think that you understand the subtleties of an encryption algorithm or to copy and paste crypto code from somewhere online, but you will generally mess up if you try to do crypto yourself.

---

7. Please disregard this if you are in fact a cryptographer.

That said, you should be aware of the CommonCrypto framework, if only so you can tell when other developers are trying to play cryptographer. There are some lower-level primitives for encryption and decryption operations, but the only one that you have any excuse for playing with is CCCrypt. Listing 13-14 shows one example of how you might use it.

```
CCCrypt(CCOperation op, CCAlgorithm alg, CCOptions options,
    const void *key, size_t keyLength, const void *iv, const void *dataIn,
    size_t dataInLength, void *dataOut, size_t dataOutAvailable,
    size_t *dataOutMoved);
```

Listing 13-14: Method signature for CCCrypt

The CCCrypt method takes 11 arguments: control over the algorithm, key length, initialization vector, operation mode, and so on. Each one is a potential place to make a cryptographic mistake. In my experience, there are several common pitfalls that developers run into with CCCrypt, which I'll describe here. Don't make the same mistakes!

Broken Algorithms to Avoid

CCCrypt supports known bad encryption algorithms, such as DES, and if you use one, your app will almost certainly be susceptible to cryptographic attacks and brute-forcing. Even if you're using the more modern AES, CCCrypt will let you switch from the default Cipher Block Chaining (CBC) mode to Electronic Code Book (ECB) using CCOptions, which is another bad idea. Using ECB mode causes identical blocks of plaintext to encrypt to identical blocks of ciphertext.[8] This is a problem because if attackers know one piece of encrypted data, they can infer the contents of other blocks. This can typically be solved with a salt or initialization vector, but they can have problems as well.

Broken Initialization Vectors

The specification for AES's CBC mode requires a nonsecret initialization vector (IV) to be supplied to the algorithm. The IV helps to randomize the encryption and produce distinct ciphertexts even if the same plaintext is encrypted multiple times. That way, you don't need to generate a new key every time to prevent disclosure of identical blocks of data.

It's important that you never reuse an IV under the same key, however, or two plaintext messages that begin with the same bytes will have ciphertext beginning with the same sequence of block values. This would reveal information about the encrypted messages to an attacker. As such, it's important to use a random IV for each cryptographic operation.

---

8. I see this *all the time*. No one should ever switch from a secure default to ECB mode, but I still come across this problem every month or two.

You should also always make sure your call to AES CBC mode encryption functions don't pass in a null initialization vector. If they do, then multiple sets of messages will be encrypted using the same key and IV, resulting in the situation I just described.

As you can see, using a static IV or a null IV has the same result: small blocks of ciphertext containing the same data will appear identical. An example of where this might be a problem would be a password manager, where encrypted keys are stored; if an attacker can read this data and determine that some of the ciphertexts are identical, they will know that the same password is used for multiple websites.

Broken Entropy

In the worst case, you may come across code that uses rand to attempt to obtain random bytes (rand being cryptographically insecure and not meant for use in cryptographic operations). The official Cocoa way to obtain entropy is via SecRandomCopyBytes.

```
uint8_t randomdata[16];
int result = SecRandomCopyBytes(kSecRandomDefault, 16, (uint8_t*)randomdata);
```

This code effectively acts as a wrapper of */dev/random*, reading entropy from the kernel's built-in entropy pool. Note that the kSecRandomDefault constant is not available on OS X, so if you're writing code to be portable, simply specify NULL as the first argument. (Under iOS, this is equivalent to using kSecRandomDefault.)

Poor Quality Keys

People often mistakenly use a user-supplied password as an encryption key. Especially on mobile devices, this results in a fairly weak, low-entropy encryption key. Sometimes, it's as bad as a four-digit PIN. When using user-supplied input to determine an encryption key, a key derivation algorithm such as PBKDF2 should be used. The CommonCrypto framework provides this with CCKeyDerivationPBKDF.

PBKDF2 is a key derivation function that uses a passphrase plus repeated iterations of a hashing algorithm to generate a suitable cryptographic key. The repeated iterations intentionally cause the routine to take longer to complete, making offline brute-force attacks against the passphrase far less feasible. CCKeyDerivationPBKDF supports the following algorithms for iterators:

- kCCPRFHmacAlgSHA1

- kCCPRFHmacAlgSHA224

- kCCPRFHmacAlgSHA256

- kCCPRFHmacAlgSHA384

- kCCPRFHmacAlgSHA512

If at all possible, you should be using at least SHA-256 or above. SHA-1 should be considered deprecated at this point because advances have been made to speed up cracking of SHA-1 hashes in recent years.

Performing Hashing Operations

In some circumstances, you may need to perform a hashing operation to determine whether two blobs of data match, without comparing the entire contents. This is frequently used to verify a file against a "known good" version or to verify sensitive information without storing the information itself. To perform a simple hashing operation on a string, you can use the CC_SHA family of methods as follows:

```
char secret[] = "swordfish";
size_t length = sizeof(secret);
unsigned char hash[CC_SHA256_DIGEST_LENGTH];
```

❶ `CC_SHA256(data, length, hash);`

This code simply defines a secret and its length and makes a char hash to contain the result of the hashing operation. At ❶, the call to CC_SHA_256 takes whatever has been put into data, calculates the hash, and stores the result in hash.

You may also be used to using OpenSSL for hashing functions. iOS does not include OpenSSL, but it does include some compatibility shims for using OpenSSL-dependent hashing code. These are defined in *CommonDigest.h*, shown in Listing 13-15.

```
#ifdef COMMON_DIGEST_FOR_OPENSSL

--snip--

#define SHA_DIGEST_LENGTH        CC_SHA1_DIGEST_LENGTH
#define SHA_CTX                  CC_SHA1_CTX
#define SHA1_Init                CC_SHA1_Init
#define SHA1_Update              CC_SHA1_Update
#define SHA1_Final               CC_SHA1_Final
```

Listing 13-15: OpenSSL compatibility hooks for CommonCrypto hashing functions

So as long as you define COMMON_DIGEST_FOR_OPENSSL, OpenSSL-style hashing operations should work transparently. You can see an example in Listing 13-16.

```
#define COMMON_DIGEST_FOR_OPENSSL
#include <CommonCrypto/CommonDigest.h>
```

```
SHA_CTX ctx;
unsigned char hash[SHA_DIGEST_LENGTH];

SHA1_Init(&ctx);
memset(hash, 0, sizeof(hash));
SHA1_Update(&ctx, "Secret chunk", 12);
SHA1_Update(&ctx, "Other secret chunk", 18);
SHA1_Final(hash, &ctx);
```

Listing 13-16: OpenSSL-style chunked SHA hashing

Listing 13-16 uses SHA1_Update and SHA1_Final, which is more appropriate for hashing a large file, where reading the file in chunks reduces overall memory usage.

Ensuring Message Authenticity with HMACs

It's important to make sure that encrypted message data hasn't been tampered with or corrupted and that it was produced by a party in possession of a secret key. You can use a keyed *Hash Message Authentication Code (HMAC)* as a mechanism to guarantee the authenticity and integrity of a message. In an iOS application, you could use this to verify the authenticity of messages sent between applications or to have a remote server verify that requests were produced by the correct application. (Just take care that the key is generated and stored in such a way that it is unique to the device and well-protected.)

To calculate an HMAC, you just need a key and some data to pass to the CCHmac function, as shown in Listing 13-17.

```
#include <CommonCrypto/CommonDigest.h>
#include <CommonCrypto/CommonHMAC.h>

❶ NSData *key = [@"key for the hash" dataUsingEncoding:NSUTF8StringEncoding];
❷ NSData *data = [@"data to be hashed" dataUsingEncoding:NSUTF8StringEncoding];
❸ NSMutableData *hash = [NSMutableData dataWithLength:CC_SHA256_DIGEST_LENGTH];
❹ CCHmac(kCCHmacAlgSHA256, [key bytes], [key length], [data bytes], [data length],
        [hash mutableBytes]);
```

Listing 13-17: Calculating an HMAC

Note that Listing 13-17 is simplified to show the basic mechanism; embedding a static key in your source code is not a recommended practice. In most cases, this key should be dynamically generated and stored in the Keychain. The operation is fairly simple. At ❶, the key for the hash is passed in as a UTF-8 encoded string (this is the part that should be stored in the Keychain). At ❷, the data to be hashed is passed in, also as a UTF-8 string. Then an NSMutableData object is constructed ❸ to store the hash for later use and all three components are passed to the CCHmac function at ❹.

Wrapping CommonCrypto with RNCryptor

If you need to use the encryption functionality exposed by CommonCrypto, RNCryptor is a good framework.[9] This wraps CommonCrypto and helps perform the most common function needed from it: encrypting data via AES with a user-supplied key. RNCryptor also helps you by providing sane defaults. The basic examples given in the instructions should be sufficient for most usage. See Listing 13-18 for basic usage.

```
NSData *data = [@"Data" dataUsingEncoding:NSUTF8StringEncoding];
NSError *error;
NSData *encryptedData = [RNEncryptor encryptData:data
                              withSettings:kRNCryptorAES256Settings
                              password:aPassword
                              error:&error];
```

Listing 13-18: Encryption with RNCryptor

Simply pass in your data to the encryptData method, along with a constant specifying the encryption settings you want to use, a key (pulled from the Keychain or from user input), and an NSError object to store the result.

Decrypting data (Listing 13-19) is more or less the inverse of encrypting, except that you do not need to provide the kRNCryptorAES256Settings constant because this is read from the header of the encrypted data.

```
NSData *decryptedData = [RNDecryptor decryptData:encryptedData
                               withPassword:aPassword
                               error:&error];
```

Listing 13-19: Decrypting RNCryptor-encrypted data

Encrypting streams or larger amounts of data while keeping memory usage sane is slightly more involved (see *https://github.com/rnapier/RNCryptor* for current examples), but the examples shown here cover the most common use case you'll likely encounter.

NOTE   *An older version of RNCryptor suffered from a vulnerability[10] that could allow an attacker to manipulate a portion of the decrypted data by altering the ciphertext, so make sure that your code is using the most up-to-date version of RNCryptor.*

---

9. *https://github.com/rnapier/RNCryptor*

10. *http://robnapier.net/blog/rncryptor-hmac-vulnerability-827*

Local Authentication: Using the TouchID

In iOS 8, Apple opened up the Local Authentication API so that third-party apps could use the fingerprint reader as an authenticator. The Local Authentication API is used by instantiating the LAContext class and passing it an authentication policy to evaluate; currently, only one policy is available, which is to identify the owner biometrically. Listing 13-20 shows this process in detail. Note that using this API doesn't give developers access to the fingerprint—it just gives a success or failure from the reader hardware.

```
❶   LAContext *context = [[LAContext alloc] init];
❷   NSError *error = nil;
❸   NSString *reason = @"We use this to verify your identity";

❹   if ([context canEvaluatePolicy:LAPolicyDeviceOwnerAuthenticationWithBiometrics
            error:&error]) {
❺       [context evaluatePolicy:LAPolicyDeviceOwnerAuthenticationWithBiometrics
                localizedReason:reason
                          reply:^(BOOL success, NSError *error) {
                          if (success) {
❻                             NSLog(@"Hooray, that's your finger!");
                          } else {
❼                     NSLog(@"Couldn't read your fingerprint. Falling back to PIN or
        something.");
                }
            }];
    } else {
        // Something went wrong. Maybe the policy can't be evaluated because the
        // device doesn't have a fingerprint reader.
❽       NSLog(@"Error: %@ %@", error, [error userInfo]);
    }
```

Listing 13-20: Authenticating the user via a fingerprint

First, this code creates an LAContext object ❶ and an NSError object ❷ to contain the results of the operation. There also needs to be a reason to present to the user when the UI asks for their fingerprint ❸. After creating these things, the code checks whether it can evaluate the LAPolicyDeviceOwner-AuthenticationWithBiometrics policy at ❹.

If evaluation is possible, then the policy is evaluated ❺; the reason and a block to handle the results of the evaluation are also passed to the evaluatePolicy method. If the fingerprint authenticates successfully, you can have the application allow whatever action it's performing to continue ❻. If the fingerprint is invalid, then depending on how you choose to write your application, it can fall back to a different method of authentication or authentication can fail entirely ❼.

If the call to `canEvaluatePolicy` at ❹ fails, then the execution ends up at ❽. This is most likely to happen if the user's device doesn't support the fingerprint reader, fingerprint functionality has been disabled, or no fingerprints have been enrolled.

How Safe Are Fingerprints?

As with most other forms of biometric authentication, fingerprint authentication is an imperfect approach. It's convenient, but given that you leave your fingerprints all over the place, it's not difficult to re-create a mold that would effectively simulate your finger. In the United States, law enforcement is legally allowed to use fingerprints to unlock devices, whereas they cannot compel someone to divulge their passcode.

There are a couple of things that developers can do to address these shortcomings. The most obvious is to provide the user with an option to use a PIN instead of using the TouchID, or perhaps in addition to the TouchID. Another thing that can help mitigate fingerprint cloning attacks is to implement a system similar to the one that Apple uses to handle the lock screen: after three unsuccessful attempts, revert to a PIN or require the user's password. Because successfully getting a cloned fingerprint is an unreliable process, this may help prevent a successful fraudulent fingerprint authentication.

Closing Thoughts

Encryption and authentication features aren't always the most straightforward to use, but given the importance of user data privacy, both from a legal and reputational standpoint, correct deployment of these features is crucial. This chapter should have given you a reasonable idea of the strategies you might encounter or need to deploy. Protecting user privacy is a broader topic than just encryption, though—you'll be turning your attention to that in the next chapter.

14

MOBILE PRIVACY CONCERNS

People tend to carry location-aware mobile devices wherever they go, and they store tons of personal data on these devices, making privacy a constant concern in mobile security. Modern iOS devices allow applications (upon request) to read people's location data, use the microphone, read contacts, access the M7 motion processor, and much more. Using these APIs responsibly not only is important to users but also can help reduce liability and increase the chances of the application being gracefully accepted into the App Store.

I discussed a fair bit of privacy-related content in Chapter 10; this was largely in regard to accidental data leakage. In this chapter, I'll cover privacy issues that affect both users and app authors when intentionally gathering and monitoring user data, as well as mitigations for some potential pitfalls.

Dangers of Unique Device Identifiers

iOS's *unique device identifiers (UDIDs)* stand as something of a cautionary tale. For most of iOS's history, the UDID was used to uniquely identify an individual iOS device, which many applications then used to track user activity or associate a user ID with particular hardware. Some companies used these identifiers as access tokens to remote services, which turned out to be a spectacularly bad idea.

Because many organizations were in possession of a device's UDID and because UDIDs weren't considered sensitive, companies that did use the UDID effectively as an authenticator were suddenly in a situation where thousands of third parties had their users' credentials. Software developers also widely assumed that the UDID was immutable, but tools had long been available to spoof UDIDs, either globally or to a specific application.

Solutions from Apple

As a result of those issues, Apple now rejects newly submitted applications that use the uniqueIdentifier API, directing developers to instead use the identifierForVendor API. This API returns an instance of the NSUUID class. The identifierForVendor mechanism should return the same UUID for all applications written by the same vendor on an iOS device, and that UUID will be backed up and restored via iTunes. It is not immutable, however, and can be reset by the user.

Older applications in the App Store that use uniqueIdentifier are returned a string starting with FFFFFFFF, followed by the string normally returned by identifierForVendor. Similarly, applications using gethostuuid are now rejected from the App Store, and existing apps receive the identifierForVendor value when calling this function.

Applications that use the NET_RT_IFLIST sysctl or the SIOCGIFCONF ioctl to read the device's MAC address now receive 02:00:00:00:00:00 instead. Of course, using a MAC address as any kind of token or authenticator has always been a terrible idea; MAC addresses leak over every network you connect to, and they're easy to change. The nonspecific return value appropriately punishes developers who have taken this approach.

For advertising and tracking purposes, Apple introduced the property advertisingIdentifier of the ASIdentifierManager class. This property returns an NSUUID that is available to all application vendors, but like uniqueIdentifier, that NSUUID can be wiped or changed (as shown in Figure 14-1).

Figure 14-1: The user interface for indicating that the advertisingIdentifier *should be used for limited purposes*

The difference between this system and the original `uniqueIdentifier` API is that `advertisingIdentifier` is explicitly

- only for advertising and tracking;
- not immutable; and
- subject to user preferences.

These aspects of `advertisingIdentifier` ostensibly give the user control over what tracking advertisers are allowed to use the mechanism for. Apple states that an application must check the value of `advertisingTrackingEnabled`, and if set to `NO`, the identifier can be used only for "frequency capping, conversion events, estimating the number of unique users, security and fraud detection, and debugging."[1] Unfortunately, that list could encompass pretty much anything advertisers want to do with the `advertisingIdentifier`.

You can determine the value of `advertisingTrackingEnabled` as shown in Listing 14-1.

```
➊ BOOL limittracking = [[ASIdentifierManager sharedManager]
       advertisingTrackingEnabled];

➋ NSUUID *advid = [[ASIdentifierManager sharedManager] advertisingIdentifier];
```

Listing 14-1: Determining whether limited ad tracking is enabled and fetching the `advertisingIdentifier`

The call to `advertisingTrackingEnabled` at ➊ reads the user preference for the advertising tracking ID before reading the `advertisingIdentifier` itself at ➋.

Rules for Working with Unique Identifiers

There are a few general rules to follow when working with unique identifiers of any type. First, never assume identifiers are immutable. Any identifier supplied by the device can be changed by someone in physical possession of the device. Second, never assume a 1:1 relationship between devices and identifiers. Identifiers can be moved from one device to another and as such cannot be trusted to uniquely identify a single device. Because identifiers can change, aren't unique, and may be widely distributed, you also shouldn't use them to authenticate users. Finally, keep identifiers as anonymous as possible. They might be useful for tracking general trends in user behavior, but don't tie an identifier to a user identity unless there's a compelling need to do so.

_____

1. *http://developer.apple.com/library/ios/#documentation/AdSupport/Reference/ASIdentifierManager_Ref/ASIdentifierManager.html*

Mobile Safari and the Do Not Track Header

Starting with iOS 6, Mobile Safari includes the option to enable the Do Not Track mechanism,[2] which tells the remote server that the user wants to opt out of being tracked by certain parties. This option is expressed with the HTTP_DNT header. When set to 1, the header indicates that the user consents to being tracked only by the site that is currently being visited. When set to 0, it indicates that the user doesn't want to be tracked by any party. Users can enable this mode in the Safari settings (Figure 14-2).

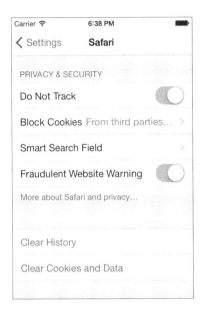

Figure 14-2: The user interface for enabling Do Not Track

At a minimum, it makes sense to assume that users want to protect details of their activity from third-party advertisers or analytics firms. This is the behavior specified by an HTTP_DNT value of 1, which is the header that iOS sends by default.

But the definition of tracking varies. The specification for the Do Not Track mechanism itself notes the following:

> The WG has not come to consensus regarding the definition of tracking and the scope of DNT. As such, a site cannot actually say with any confidence whether or not it is tracking, let alone describe the finer details in a tracking status resource.[3]

---

2. *http://www.w3.org/TR/tracking-dnt/*

3. *http://www.w3.org/2011/tracking-protection/drafts/tracking-dnt.html*

According to the specification, websites can prompt the user to opt into specific tracking scenarios using the storeSiteSpecificTrackingException JavaScript API, but this functionality is not widely implemented at the time of this writing.

Cookie Acceptance Policy

Cookies on iOS are managed via the NSHTTPCookieStorage API. The method sharedHTTPCookieStorage returns the cookie store, but despite the method's name, iOS cookie storage is specific to each application. Cookies actually live in a database under the application main bundle directory.

NOTE   *The name* sharedHTTPCookieStorage *originates from OS X, where the OS uses a global cookie store shared among all applications.*

Cookies used by the URL loading system are accepted according to a systemwide shared cookieAcceptPolicy, which any application can specify. This policy can be set to any of the following:

NSHTTPCookieAcceptPolicyAlways   Accept and store all received cookies. This is the default.

NSHTTPCookieAcceptPolicyOnlyFromMainDocumentDomain   Accept only first-party cookies.

NSHTTPCookieAcceptPolicyNever   Never accept cookies.

Note that on devices running anything older than iOS 7, the cookie acceptance policy is shared among applications, which could cause problems for your application. On such devices, when another running application changes its acceptance policy, your app's policy changes as well. For example, an application that relies on third-party cookies for advertising revenue might repeatedly set its cookie policy to NSHTTPCookieAcceptPolicyAlways, changing yours to the same policy in the process.

Fortunately, you can specify your preferred cookieAcceptPolicy using events such as didFinishLaunchingWithOptions, and you can monitor for changes to the cookie acceptance policy while your program is running, as shown in Listing 14-2.

```
❶ [[NSNotificationCenter defaultCenter] addObserver:self selector:@selector
       (cookieNotificationHandler:)
     name:NSHTTPCookieManagerAcceptPolicyChangedNotification object:nil];


- (void) cookieNotificationHandler:(NSNotification *)notification {
    NSHTTPCookieStorage *cookieStorage = [NSHTTPCookieStorage
     sharedHTTPCookieStorage];
```

❷   ```
      [cookieStorage setCookieAcceptPolicy:
          NSHTTPCookieAcceptPolicyOnlyFromMainDocumentDomain];
  }
```

---

*Listing 14-2: Registering to receive notifications when the cookie acceptance policy changes*

Listing 14-2 registers an `NSNotificationCenter` at ❶, which listens for `NSHTTPCookieManagerAcceptPolicyChangedNotification`. If the policy changes, the selector identified at ❶, `cookieNotificationHandler`, will be called. In the `cookieNotificationHandler`, you set the policy to `NSHTTPCookieAcceptPolicyOnly-FromMainDocumentDomain` at ❷.

In iOS 7 and later, changes in cookie management policy affect only the running application. Applications can also specify different cookie management policies for different HTTP sessions via `NSURLSession`. For more on this, see "Using NSURLSession" on page 117.

## Monitoring Location and Movement

One of the most useful features of mobile platforms is their ability to make information and functionality relevant to a user's current physical location and method of movement. iOS devices primarily determine location based on Wi-Fi and GPS, and they monitor body movement with the M7 motion processor.

Gathering location and movement data has dangers, however. In this section, I'll discuss how gathering both types of data works and why you should take care when storing such information.

### How Geolocation Works

Wi-Fi geolocation scans for available wireless access points and queries a database that has a record of access points and their GPS coordinates. These databases are built by third parties that effectively wardrive entire cities and note the coordinates of each discovered access point. Of course, this can result in inaccurate results in some circumstances. For example, if someone travels with network equipment, or relocates it, the location data may not get updated for some time.

GPS can provide more specific navigation information, as well as motion information, to track users in transit. This requires the ability to contact GPS satellites, which is not always possible, so GPS is often used as a fallback or when a high degree of accuracy is required. GPS is also required to determine information such as speed or altitude.

### The Risks of Storing Location Data

Few aspects of mobile privacy have generated as much negative press as tracking users via geolocation data. While useful for an array of location-
```

aware services, a number of issues arise when location data is recorded and stored over time. Most obvious are privacy concerns: users may object to their location data being stored long-term and correlated with other personal information.[4] Aside from PR concerns, some European countries have strict privacy and data protection laws, which must be taken into account.

A less obvious problem is that storing location data linked to specific users could leave you legally vulnerable. When you store location data along with data that links it to a specific individual, that data could be subpoenaed by law enforcement or litigators. This often occurs in divorce cases, where lawyers attempt to demonstrate infidelity by showing the physical comings and goings of one of the parties in the course of a relationship; toll authorities that use electronic tracking have had to respond to these inquiries for years.

Restricting Location Accuracy

Because precise historical location data raises such privacy and liability concerns, it's important to use the least degree of accuracy necessary for your intended purpose. For example, if your application is designed to determine what city or neighborhood a user is in for the purpose of making a dinner reservation, you'll only need to get a user's location within a kilometer or so. If your purpose is to find the nearest ATM to a user, you'll want to use something significantly narrower. The following are the geolocation accuracy constants available via the Core Location API:

- kCLLocationAccuracyBestForNavigation
- kCLLocationAccuracyBest
- kCLLocationAccuracyNearestTenMeters
- kCLLocationAccuracyHundredMeters
- kCLLocationAccuracyKilometer
- kCLLocationAccuracyThreeKilometers

Restricting location accuracy to the least degree necessary is not only a best practice for privacy and legal reasons but also reduces power consumption. This is because less accurate methods use the rather quick Wi-Fi detection mechanisms and update less frequently, while the highest accuracy settings will often use GPS and update frequently.

If you do need to store multiple instances of a user's location over time, ensure that procedures are in place to prune this data eventually. For instance, if you need to reference only a month's worth of location data at a time, ensure that older data is properly sanitized or erased. If you're using location data for analytics that don't require linking to a specific user, omit or remove any data that uniquely identifies the user.

---

4. *http://www.pskl.us/wp/wp-content/uploads/2010/09/iPhone-Applications-Privacy-Issues.pdf*

Requesting Location Data

Permission data is requested using `CLLocationManager`, which specifies an accuracy constant as well as whether your app needs location data when it's backgrounded. Listing 14-3 shows an example invocation.

```
❶  [self setLocationManager:[[CLLocationManager alloc] init]];
❷  [[self locationManager] setDelegate:self];
❸  [[self locationManager] setDesiredAccuracy:kCLLocationAccuracyHundredMeters];

   if ([[self locationManager] respondsToSelector:
       @selector(requestWhenInUseAuthorization)]) {
❹      [[UIApplication sharedApplication] sendAction:
       @selector(requestWhenInUseAuthorization)
                                           to:[self locationManager]
                                         from:self
                                     forEvent:nil];
❺      [[self locationManager] startUpdatingLocation];
   }
```

Listing 14-3: Requesting location data permissions

Here, a `CLLocationManager` is allocated ❶ and its delegate is set to `self` ❷. Then the desired accuracy of about 100 meters is set at ❸. At ❹, the permission request is sent, which will cause the authorization prompt to appear to the user. Finally, at ❺, there's a request for the manager to start monitoring the user's location.

Note that as of iOS 8, the location manager won't actually start unless you have a description of why you need location data. This is specified in your *Info.plist* file, using either `NSLocationWhenInUseUsageDescription` if you need to access location data only when the app is in use or `NSLocationAlwaysUsage-Description` if you'll also need to get location information from the background. Add one of these to your plist file, along with a concise but specific rationale to be displayed to the user when they're prompted to grant permission to location data.

Managing Health and Motion Information

Some of the most sensitive information that applications can handle is health information about the user. On iOS, this data can be retrieved using the HealthKit API and the APIs provided by the device's M7 motion processor, if it has one. You'll take a brief look at how to read and write this data and how to request the minimum privileges necessary for an app to function.

NOTE   *As of iOS 9, HealthKit is available only on iPhones, not on iPads.*

Reading and Writing Data from HealthKit

HealthKit information can be requested either for reading or for both reading and writing (somewhat confusingly called *sharing* by Apple). In keeping with requesting only the permissions that are absolutely necessary, request read-only access if possible. Listing 14-4 shows how permissions for specific health data are requested.

```
if ([HKHealthStore isHealthDataAvailable]) {
    HKHealthStore *healthStore = [[HKHealthStore alloc] init];
❶  HKObjectType *heartRate = [HKObjectType quantityTypeForIdentifier:
      HKQuantityTypeIdentifierHeartRate];
❷  HKObjectType *dob = [HKObjectType characteristicTypeForIdentifier:
      HKCharacteristicTypeIdentifierDateOfBirth];
❸  [healthStore requestAuthorizationToShareTypes:
    [NSSet setWithObject:heartRate]
                readTypes:[NSSet setWithObject:dob]
❹              completion:^(BOOL success, NSError *error) {
                                if (!success) {
                                    // Failure and sadness

                                } else {
                                    // We succeeded!
                                }
                }];
}
```

Listing 14-4: Requesting health data permissions

At ❶ and ❷, you specify two types of data that you'd like to access, namely, heart rate and date of birth. At ❸, you request authorization to access these, with a `completion` block to handle success or failure. Note that the `requestAuthorizationToShareTypes` is requesting read/write access, presumably because this application is intended to track and record the user's heart rate. The `readTypes` parameter specifies that you want to monitor the user's heart rate but not write to it. In this case, you're requesting the user's date of birth (something rather unlikely to change) to infer their age. Finally, to allow you to distribute the application, you'll need to enable the HealthKit entitlement in Xcode, as shown in Figure 14-3.

While HealthKit shows how to record steps, but there are more detailed ways to get motion data to help guess exactly what kind of activity the user is engaged in. This more detailed data can be retrieved using the M7 motion tracker.

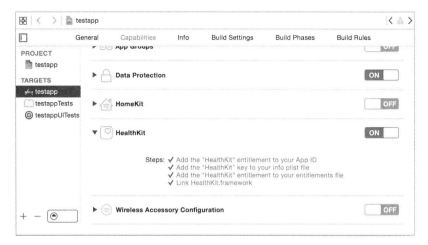

Figure 14-3: Enabling the HealthKit entitlement in Xcode

The M7 Motion Processor

The iPhone 5s introduced the M7 motion-tracking processor, which allows for the recording of detailed information about small movements while reducing the battery drain that this has incurred in the past. Fitness applications could use this data to determine the type of physical activity the user is currently engaging in and how many steps they've taken. Applications that monitor sleep quality could also take advantage of this to determine how deep the user is sleeping based upon slight movements. Obviously, the ability to determine when a user is asleep and what they're doing outside of using the phone is a significant responsibility. Apple details the degree to which users can be tracked via the M7 as follows:

> M7 knows when you're walking, running, or even driving. For example, Maps switches from driving to walking turn-by-turn navigation if, say, you park and continue on foot. Since M7 can tell when you're in a moving vehicle, iPhone 5s won't ask you to join Wi-Fi networks you pass by. And if your phone hasn't moved for a while, like when you're asleep, M7 reduces network pinging to spare your battery.[5]

Use of the M7 processor is granted in a manner similar to basic geolocation permissions. But M7 has a quirk not present in other geolocation data access: applications have access to data that was recorded before they were granted permission to access location data. If you're going to use this historic data, inform the user in your permissions message and, ideally, give them a choice as to whether to use or disregard this data.

---

5. *http://www.apple.com/iphone-5s/features/*

Requesting Permission to Collect Data

When attempting to access sensitive data such as a user's contacts, calendar, reminders, microphone, or motion data, the user will be prompted with an alert to grant or deny this access. To ensure that the user is presented with useful information as to why you need this access, define strings to be delivered to the user as part of the access prompt. Make these explanations simple but descriptive, as in Figure 14-4.

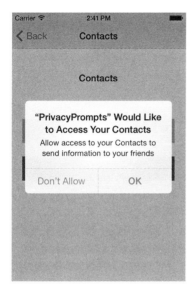

Figure 14-4: Delivering the request to the user

You can set those messages in your app's *Info.plist* file through Xcode, shown in Figure 14-5.

Application requires iPhone environment	Boolean	YES	
Privacy – Calendars Usage Description	String	Needed to silence alarms during scheduled events	
Privacy – Contacts Usage Description	String	Allow access to your Contacts to send information to your friends	
Privacy – Microphone Usage Description	String	Take voice memos during your wander	
Privacy – Motion Usage Description	String	Allow detailed motion information to track the number of steps you've taken	
Main storyboard file base name	String	MainStoryboard	
▶ Required device capabilities	Array	(1 item)	
▶ Supported interface orientations	Array	(1 item)	

Figure 14-5: Describing needs for various kinds of access in an Info.plist

Additionally, ensure that your application handles the refusal of these permissions gracefully. Unlike Android, where permission granting is an all-or-nothing affair, iOS applications are expected by Apple to be able to handle having some permissions granted and others refused.

Proximity Tracking with iBeacons

Apple's iBeacons are designed to measure your proximity to hardware and take certain actions when you're within range. For example, an app could use the beacons to track your movements through a mall or store, or indicate that the car that just pulled up next to you is the Uber car you requested. iBeacon functionality is part of the Core Location API, which uses Bluetooth Low Energy (BTLE) on compatible devices to manage proximity monitoring.

In this section, I'll first discuss how some applications check for iBeacons and how iOS devices can become iBeacons. I'll end on privacy issues you should consider when using iBeacons in your own apps.

Monitoring for iBeacons

Monitoring for iBeacons is accomplished by instantiating a Core Location `CLLocationManager` and passing a `CLBeaconRegion` to the manager's `startMonitoringForRegion` method, as in Listing 14-5.

```
❶  NSUUID *myUUID = [[NSUUID alloc] initWithUUIDString:
       @"CE7B5250-C6DD-4522-A4EC-7108BCF3F7A4"];
   NSString *myName = @"My test beacon";

   CLLocationManager *myManager = [[CLLocationManager alloc] init];
   [myManager setDelegate:self];

❷  CLBeaconRegion *region = [[CLBeaconRegion alloc] initWithProximityUUID:myUUID
❸                                                     identifier:myName];
   [myManager startMonitoringForRegion:region];
```

Listing 14-5: Initiating monitoring for a specific region defined by a UUID

The NSUUID generated at ❶ is assigned to a `CLBeaconRegion` ❷ and will be used to uniquely identify that beacon. The identifier ❸ will specify the symbolic name for the beacon. Note you can register to monitor for multiple regions with the same `CLLocationManager`.

NOTE   *You can use the uuidgen(1) command in the terminal to generate a unique UUID to use as a beacon identifier.*

You'll also need to implement a `locationManager` delegate method, as in Listing 14-6, to handle location updates.

```
- (void)locationManager:(CLLocationManager *)manager
       didEnterRegion:(CLRegion *)region {
  if ([[region identifier] isEqualToString:myName]) {
```

```
    [self startRangingBeaconsInRegion:region];
  }
}
```

Listing 14-6: An example locationManager *delegate method*

This method will be called whenever a device running your application enters an iBeacon's registered region; your app can then perform whatever logic is appropriate upon entering that region. Once the application gets the notification that the device has entered the range of a beacon, it can start *ranging*, or measuring the distance between the device and the beacon.

After an application has begun ranging a beacon, the locationManager:did-RangeBeacons:inRegion delegate method (Listing 14-7) will be called periodically, allowing the application to make decisions based on the proximity of the beacon.

```
- (void)locationManager:(CLLocationManager *)manager didRangeBeacons:
    (NSArray *)beacons inRegion:(CLBeaconRegion *)region
{
  CLBeacon *beacon = [beacons objectAtIndex:0];

  switch ([beacon proximity]) {
    case CLProximityImmediate:
      //
      break;
    case CLProximityNear:
      //
      break;
    case CLProximityFar:
      //
      break;
    case CLProximityUnknown:
      //
      break;
  }
}
```

Listing 14-7: The locationManager *callback for examining beacons*

There are four constants representing proximity: CLProximityImmediate, CLProximityNear, CLProximityFar, and CLProximityUnknown. See Table 14-1 for the meanings of these values.

Table 14-1: Region Proximity (Range) Measurements

Item class	Meaning
CLProximityUnknown	The range is undetermined.
CLProximityImmediate	The device is right next to the beacon.
CLProximityNear	The device is within a few meters of the beacon.
CLProximityFar	The device is within range but near the edge of the region.

Turning an iOS Device into an iBeacon

BTLE iOS devices can also act as iBeacons, broadcasting their presence to the outside world, which can be used to detect and measure proximity between iOS devices. This is done via the CoreBluetooth framework, using a CBPeripheralManager instance and giving it a CLBeaconRegion with a chosen UUID (Listing 14-8).

```
❶ NSUUID *myUUID = [[NSUUID alloc] initWithUUIDString:
       @"CE7B5250-C6DD-4522-A4EC-7108BCF3F7A4"];
❷ NSString *myName = @"My test beacon";

❸ CLBeaconRegion *region = [[CLBeaconRegion alloc] initWithProximityUUID:myUUID
                                                identifier:myName];

❹ NSDictionary *peripheralData = [region peripheralDataWithMeasuredPower:nil];

❺ CBPeripheralManager *manager = [[CBPeripheralManager alloc] initWithDelegate:self
                                                queue:nil];

❻ [manager startAdvertising:peripheralData];
```

Listing 14-8: Turning your application into an iBeacon

The code generates a UUID at ❶ and a symbolic name at ❷, and then defines a region at ❸. At ❹, the peripheralDataWithMeasuredPower method returns a dictionary with the information needed to advertise the beacon (the nil parameter just tells the code to use the default signal strength parameters for the device). At ❺, an instance of CBPeripheralManager is instantiated and finally the peripheralData ❻ is passed to the manager so it can begin advertising.

Now that you've taken a look at how iBeacons are managed, let's look at some of the privacy implications of implementing them.

iBeacon Considerations

Obviously, iBeacons provide extremely detailed information about a user's whereabouts. Beacons don't have to be dumb transmitters; they can also be programmable devices or other iOS devices that can record location updates and deliver them to central servers. Users are likely to object to this data being tracked over the long term, so as with other geolocation data, avoid keeping any beacon logs for any longer than they're specifically needed. Also, don't tie the time and beacon information in such a way that they can be correlated with a specific user in the long term.

Your app should turn the device it's installed on into a beacon sparingly. Becoming a beacon makes the device discoverable, so be sure to inform the user of your intentions in a manner that communicates that fact. If possible, perform Bluetooth advertising for only a limited time window, ceasing it once necessary data has been exchanged.

Now that you've looked at some of the many ways apps gather information about users, let's look at some of the policy guidelines that will dictate how those apps handle personal data.

Establishing Privacy Policies

For your own protection, always explicitly state a privacy policy in your application. If your app is set to Made for Kids, a privacy policy is both an App Store requirement and a legal one, as required by the Children's Online Privacy Protection Act (COPPA).

I'm no lawyer, so of course, I can't give specific legal advice on how your policy should be implemented. However, I would advise you to include the following in your privacy policy:

- The information your app gathers and whether it is identifying or non-identifying (that is, whether it can be tied back to a specific user)
- The mechanisms by which information is gathered
- The reasons for gathering each type of data
- How that data is processed and stored
- The retention policy of the data (that is, how long data is stored)
- If and how the information you gather is shared with third parties
- How users can change data collection settings if desired
- Security mechanisms in place to protect user data
- A history of changes to the privacy policy

The Electronic Frontier Foundation (EFF) provides a good template for developing an effective and informative privacy policy, which you can find at *https://www.eff.org/policy*.

Do note that Apple has some specific requirements for how to implement privacy policies in the application and how they should be made available. Specifically, all applications that offer autorenewed or free subscriptions and apps that are categorized as Made for Kids must include a URL to a privacy policy. If the application is set to Made for Kids, the policy needs to be localized for each localization within the application.[6]

Closing Thoughts

In light of disclosures about massive government surveillance in the United States and abroad, consumer awareness and concern about companies gathering and correlating their personal information and habits is likely to increase. It's also become clear that the more information you gather on your users, the greater your company risks exposure. Companies with the most detailed information on their users are those most attractive to government intrusion, either by subpoena, monitoring, or active hacking by intelligence agencies.

In summary, always clearly define your intentions and minimize data gathered to limit your exposure and to build and maintain trust with consumers.

---

6. *https://developer.apple.com/library/ios/documentation/LanguagesUtilities/Conceptual/iTunesConnect_Guide/8_AddingNewApps/AddingNewApps.html*

INDEX

A

Address Sanitizer (ASan), 55
Address Space Layout Randomization
    (ASLR), 8, 53–54, 87
advertisingIdentifier, 235
advertisingTrackingEnabled, 235
AES algorithm, 226–227
AFNetworking, 122–124
alloc, 19
.app directory, 78–79
Apple System Log (ASL), 161–164
application anatomy, 27–38
    Bundle directory, 33–34
    Data directory, 34–37
    device directories, 32–33
    Documents directory, 34–35
    Shared directory, 37
applicationDidEnterBackground, 20, 167,
    179–180, 183
application extensions, 140–144
    extensionPointIdentifier, 144
    extension points, 140
    NSExtensionActivationRule, 142
    NSExtensionContext, 143
    NSExtensionItem, 143
    shouldAllowExtensionPoint
        -Identifier, 143
    third-party keyboards, 144
Application Support directory, 35
applicationWillResignActive, 180
applicationWillTerminate, 20, 167, 183
app review, 3–4, 10–12
    evading, 11–12
App Store, 3–4,
    review process, 10–12
        bypassing, 11–12
ARC (Automatic Reference
    Counting), 19
ASan (Address Sanitizer), 55
ASIHTTPRequest, 122, 124–125
ASL (Apple System Log), 161–164

ASLR

ASLR (Address Space Layout
    Randomization), 8, 53–54, 87
authentication
    biometrics, 231–232
    fingerprint authentication,
        safety of, 232
    HTTP basic authentication,
        110–111, 119–121
    Local Authentication API, 231–232
    TouchID, 231–232
        LAContext, 231–232
autocorrection, 175–177
Automatic Reference Counting
    (ARC), 19
autoreleasepool, 19

B

backtrace (bt) command, 65–66
BEAST attack, 117
biometrics, 231–232
black-box testing, 77
blacklisting, 200
blocks, Objective-C
    declaring, 18
    exposing to JavaScript, 150–151
Bluetooth Low Energy (BTLE), 244
Bluetooth PAN, 125
Bonjour, 125
Boot ROM, 4
breakpoints, 62
    actions, 70–72, 164
    conditions, 72
    enabling/disabling, 64
    setting, 62–64
brute-forcing, PINs, 214
bt (backtrace) command, 65–66
BTLE (Bluetooth Low Energy), 244
buffer overflows, 12, 193–196
    example, 194–195
    preventing, 195–196
Bundle directory, 33–34
bundle ID, 33, 138

bundle seed ID, 218–219
BurpSuite, 43–47

C

CA (certificate authority), 114–115
CA certificate, 44
  certificate management, 47
  certificate pinning, 114–117, 124
    defeating, 96–97
cache management, 170–171
  removing cached data, 171–174
Caches directory, 35–36
caching, 36
CALayer, 182–183
canonical name (CN), 128–129
canPerformAction, 168–169
categories, Objective-C, 22–23
CBPeripheralManager, 246
CCCrypt, 186, 226
CCHmac, 229
CCRespring, 79
certificate authority (CA), 114–115
certificateHandler, 126
CFArrayRef, 112–113
CFBundleURLSchemes, 133
CFDataRef, 113
CFPreferences, 36, 178
CFStream, 48, 107, 128–129
chflags command, 42
clang, 51–53
class-dump command, 90, 92–93
CLBeaconRegion, 244, 246
CLLocationManager, 240, 244
CN (canonical name), 128–129
Cocoa, 14
Cocoa Touch, 14
code segment, 193–195
codesign command, 82
CommonCrypto, 151, 230, 230
  CCCrypt, 226
CompleteUnlessOpen, 222–223
CompleteUntilFirstUser-
  Authentication, 220
cookieAcceptPolicy, 237
cookies, 36
  acceptance policy, 237–238
  theft of, 114
Cookies directory, 36
copy/paste, disabling, 168–169

Cordova, 150, 154–157
Cordova.plist, 156
CoreBluetooth, 246
Core Data, 204, 223
CRIME attack, 118
cross-site scripting (XSS), 199–200
  input sanitization, 200–201
  output encoding, 201–202
cryptid, 81, 86–90
cryptoff, 86–90
cryptsize, 86–90
cURL, 78
  certificates, 93–94
Cycript, 90, 93–94
Cydia, 31, 77
Cydia Substrate, 78, 97–100

D

DAC (discretionary access
    control), 4–5
Data directory, 34–37
data leakage, 161–188
  Apple System Log, 161–164
  autocorrection, 175–177
  breakpoint actions, 164
  cache management, 170–174
  dynamic-text.dat, 177
  ephemeral sessions, 173
  HTTP caches, 169–174
    local storage, 174
  iCloud, 161
    avoidance of, 188
  keylogger, 177
  NSLog, 161–164
    disabling, 163
  NSURLSession, 173
  pasteboards, 164–169
    canPerformAction, 168
    disabling copy/paste, 168–169
    pasteboardDB file, 165–167
    pasteboardWithUniqueName, 165–167
    protecting data, 167–169
    UISearchBar, 165
    wiping, 167
  snapshots, 178–184
    applicationDidEnterBackground,
      179–180
    preventing suspension, 183–184
    screen obfuscation, 179–183

state preservation, 184–187
  restorationIdentifier, 184–185
user preferences, 178
Data Protection API, 7–8, 219–225
  Class Key, 220
  CompleteUnlessOpen, 222–223
  CompleteUntilFirstUser-
    Authentication, 220
  DataProtectionClass, 223
  Data Protection entitlement,
    223–224
  delegate methods, 224
  detecting, 225
  FileProtectionComplete, 220–221
  isProtectedDataAvailable, 225
  protection levels, 220–223
DataProtectionClass entitlement, 157
data segment, 193–195
dataTaskWithRequest, 18
data theft, 161
dd command, 88
debugging, 61–75
  breakpoints, 62
    actions, 70–72
    conditions, 72
    enabling/disabling, 64
    setting, 62–64
  Debug Navigator, 65
  debugserver, 81–84
    connecting to, 83
    installing, 81–82
  fault injection, 72–73
  frames and variables, 64–68
  lldb, 62–75
    backtrace (bt) command, 65–66
    expr command, 69
    frame select command, 66–67
    frame variable command, 66
    image list, 87
    print object command, 67–68
  object inspection, 68
  tracing data, 74
  variables and properties, 69–70
Debug Navigator, 65
debugserver, 81–84
  connecting to, 83
  installing, 81–82
decodeRestorableStateWithCoder, 184
decrypting binaries, 80–90

.default_created.plist, 32
Default-Portrait.png, 179
defaults command, 42
delegation, 20
DES algorithm, 226
deserialization, 21
developer team ID, 138
device directories, 32–33
Device Key, 7
device.plist, 32
didFinishNavigation, 159–160
did messages, 20
didReceiveCertificate, 126
disassembly, with Hopper, 94–96
discretionary access control
    (DAC), 4–5
Documents directory, 34–35
Do Not Track, 236–237
dpkg command, 96, 99–101
DTrace, 55, 61
dumpdecrypted command, 80
_dyld_get_image_name, 10
dylibs, 10
dynamic analysis, 55
dynamic patching, 11–12

E

emulator, see Simulator
encodewithcoder, 21–22
encrypted segment, 84–90
encryption, 211–230
  AES, CCB mode, 226–227
  bad algorithms, 226
  CommonCrypto, 225, 230
    CCCrypt, 226
  Curve25519, 222
  Data Protection API, 5, 7–8,
    219–225
    Class Key, 220
    CompleteUnlessOpen, 222–223
    CompleteUntilFirstUser-
      Authentication, 220
    DataProtectionClass, 223
    Data Protection entitlement,
      223–224
    delegate methods, 224
    detecting, 225
    FileProtectionComplete, 220–221

encryption, Data Protection API,
    continued
    FileProtectionCompleteUnless-
        Open, 222
    isProtectedDataAvailable, 225
    protection levels, 220–223
DES algorithm, 226
Device Key, 7
disk encryption, 5–7
Elliptic Curve Diffie-Hellman
    algorithm, 222
entropy, 227
File Key, 7
full disk encryption, 5–7
hashing, 228–230
HMAC (Hash Message
    Authentication Code),
    229–230
initialization vector (IV), 226–227
Keychain, 6–7, 113, 186, 211–219
    API, 7
    backups, 212
    iCloud synchronization, 219
    item classes, 214
    key hierarchy, 6–7
    kSecAttrAccessGroup, 218–219
    protection attributes, 212–214
    SecItemAdd, 219
    shared Keychains, 218–219
    usage, 214–217
    wrappers, 217–218
key derivation, 227–228
key quality, 227–228
Lockbox, 217
OpenSSL, 228–229
RNCryptor, 230
SecRandomCopyBytes, 227
TLS (Transport Layer Security),
    127–129
entitlements, 218, 223
entitlements.plist, 81–82
entropy, 227
Erica Utilities, 31, 78
/etc/hosts, 49
EXC_BAD_ACCESS, 191
eXecute Never (XN), 8–9
expr command, 69
extensionPointIdentifier, 144
extractIdentityAndTrust, 112–113

F

fault injection, 72–73
File Juicer, 169, 174
File Key, 7
FileProtectionComplete, 220–221
filesystem monitoring, 58–59
Finder, 42
fingerprint authentication,
    safety of, 232
forensic attackers, 161
format string attacks, 190–193
    NSString, 192–193
    preventing, 191–193
Foundation classes, 14
frames and variables, 68
frame select command, 66–67
frame variable command, 66
Full Disk Encryption, 5–7
fuzzing, 55

G

garbage collection, 18
gdb, 62
geolocation, 238
    accuracy, 239
    CLLocationManager, 240
    risks, 238–239
get-task-allow, 82
Google Toolbox for Mac, 202
GPS, 238

H

handleOpenURL, 136
hashing, 228–230
Hash Message Authentication Code
    (HMAC), 229
hasOnlySecureContent, 159–160
HealthKit, 240–241
heap, 8, 53–54, 193
hidden files, 41–42
HMAC (Hash Message Authentication
    Code), 229
Homebrew, 46, 88, 94, 99
hooking
    with Cydia Substrate, 97–100
    with Introspy, 100–103
Hopper, 94–96

HTML entities, 201
    encoding, *see* output encoding
HTTP basic authentication, 110–111,
        119–121
HTTP local storage, 174
HTTP redirects, 113–114

I

iBeacons, 244–247
    CBPeripheralManager, 246
    CLBeaconRegion, 244–246
    CLLocationManager, 244
    startMonitoringForRegion, 244
iBoot, 4
iCloud, 35, 111, 161, 212, 219
    avoidance of, 187
IDA Pro, 94
identifierForVendor, 234
iExplorer, 28–29
iGoat, 178
image list, 87
implementation, declaring, 16–17
Info.plist, 33
init, 19
initialization vector (IV), 226–227
initWithCoder, 21–22
initWithContentsOfURL, 206
injection attacks, 199–207
    cross-site scripting (XSS), 199–202
        input sanitization, 200–201
        output encoding, 200–202
    displaying untrusted data, 202
    predicate injection, 204–205
    SQL injection, 203–204
        parameterized SQL, 203–204
        SQLite, 203–204
    XML injection, 207
        XML external entities, 205–206
        XPath, 207
input sanitization, 200–201
installipa command, 80
InstaStock, 12
Instruments, 55–57
integer overflow, 196–198
    example, 197–198
    preventing, 198
interface, declaring, 15–16
interprocess communication, *see* IPC
        (interprocess communication)

Introspy, 100–103
iOS-targeted web apps, 147–160
IPA Installer Console, 78
.ipa packages, 80
IPC (interprocess communication),
        131–145
    application extensions, 131, 140
        extensionPointIdentifier, 144
        extension points, 140
        isContentValid, 143
        NSExtensionActivationRule, 142
        NSExtensionContext, 143
        NSExtensionItem, 143
        shouldAllowExtensionPoint
            -Identifier, 143
        third-party keyboards, 143–144
    canOpenURL, 138
    handleOpenURL, 136
    isContentValid, 143
    openURL, 132–137
    sourceApplication, 136
    UIActivity, 139–140
    UIPasteboard, 144
    universal links, 137–138
    URL schemes, 132–133
        CFBundleURLSchemes, 133
        defining, 132–133
        hijacking, 136–137
        validating URLs and senders, 134
iproxy command, 84
isContentValid, 143
IV (initialization vector), 226–227
ivars, 15–17, 91

J

jailbreak detection, 9–10
    futility of, 9
jailbreaking, 4, 9–10, 77
JavaScript, 11
    executing in Cordova, 154–157
    executing in UIWebView, 149–150
    stringByEvaluatingJavaScriptFrom-
        String, 149–150
JavaScript–Cocoa bridging, 150–157
JavaScriptCore, 150–154
    blocks, 150–151
    JSContext, 152–154
    JSExport, 151–152
Jekyll, 12

just-in-time (JIT) compiler, 8–9, 149
JRSwizzle, 25
JSContext, 152–154
JSExport, 151–152

K

kCFStreamSSLLevel, 129
Keychain API, 6–7, 113, 186, 211
  backups, 212
  iCloud synchronization, 219
  kSecAttrAccessGroup, 218–219
  protection attributes, 212–214
  SecItemAdd, 214–215, 219
  SecItemCopyMatching, 216
  SecItemDelete, 216
  SecItemUpdate, 215
  shared Keychains, 218–219
  usage, 214–217
  wrappers, 217–218
key derivation, 227–228
keylogging, 175–177
killall command, 79, 101
kSecAttrAccessGroup, 218–219
kSecAttrAccessible, 220
kSecAttrSynchronizable, 219

L

LAContext, 231–232
ldid command, 97
LDID (link identity editor), 97
legacy issues, from C, 189–198
  buffer overflows, 193–196
    example, 194–195
    preventing, 195–196
  format string attacks, 190–193
    NSString, 192–193
    preventing, 191–193
  integer overflow, 196–198
    example, 197–198
    preventing, 198
libc, 8
Library directory, 35–37
  Application Support directory, 35
  Caches directory, 35–36, 187
    Snapshots directory, 36
  Cookies directory, 36
  Preferences directory, 36
  Saved Application State directory, 37
LIKE operator, 205

link identity editor (LDID), 97
lipo command, 78, 85
lldb, 62–81, 83–84, 191
  backtrace (bt) command, 65–66
  breakpoints, 62
    actions, 70–72, 164
    conditions, 72
    enabling/disabling, 64
    setting, 62–64
  expr command, 69
  frame select command, 66–67
  frame variable command, 66
  image list, 87
  print object command, 67–68
llvm, 90
Local Authentication API, 231–232
Logos, 98
loopback interface, 46–47
Lua, 12

M

M7 processor, 242
MAC (mandatory access control), 4–5
MAC address, 234
Mach-O binary format, 77, 85
MachOView, 88
MacPorts, 94
malloc, 197–198
mandatory access control (MAC), 4–5
MATCHES operator, 205
MCEncryptionNone, 126
MCEncryptionOptional, 126
MCEncryptionRequired, 126
MCSession, 126
message passing, 13–15
method swizzling, 23–25
Mobile Safari, 44
MobileTerminal, 78
multipeer connectivity, 125–127
  certificateHandler, 126
  didReceiveCertificate, 126
  encryption, 125–127

N

netcat command, 78
networking, 107–129
  AFNetworking, 122–124
    certificate pinning, 123–124

ASIHTTPRequest, 122, 124–125
backgroundSessionConfiguration, 117
CFStream, 48, 107, 128–129
ephemeralSessionConfiguration, 117
multipeer connectivity, 125–127
  certificateHandler, 126
  didReceiveCertificate, 126
  encryption, 125–127
NSInputStream, 49
NSOutputStream, 49
NSStream, 48, 107, 127–128
NSURLSession, 122
URL loading system, 107–122
  HTTP basic authentication,
    110–111, 119–121
  HTTP redirects, 113–114
  NSURLConnection, 48, 108
  NSURLConnectionDelegate, 109
  NSURLCredential, 120
  NSURLCredentialStorage, 110–111
  NSURLRequest, 108
  NSURLResponse, 108
  NSURLSession, 48, 117
  NSURLSessionConfiguration,
    120–121
  NSURLSessionTaskDelegate, 119
  sharedCredentialStorage,
    120–122
  stored URL credentials, 121–122
Notification Center, 224–225
NSCoder, 185–187
NSCoding, 21–22
NSData, 113
NSExtensionContext, 143
NSExtensionItem, 143
NSFileManager, 221–223
NSHTTPCookieStorage, 237
NSHTTPRequest, 122
NSInputStream, 49
NSLog, 95, 161–164, 192
  disabling, 163
NSNotificationCenter, 224–225
NSOperation, 122
NSOutputStream, 49
NSPredicate, 204–205
NSStream, 48, 107, 127–128
NSString, 192–193, 195, 202
NSURAuthenticationMethodClient-
    Certificate, 112

NSURL, 188
NSURLCache, 74–75, 150
NSURLConnection, 48, 108, 117
NSURLConnectionDelegate, 109, 114
NSURLCredential, 113, 120
NSURLCredentialStorage, 110–111, 121
NSURLIsExcludedFromBackupKey, 35,
    187–188
NSURLProtectionSpace, 109–111, 122
NSURLProtocol, 155
NSURLRequest, 108, 148–149
NSURLResponse, 108
NSURLSession, 48, , 117–122
NSURLSessionConfiguration, 117–119
NSURLSessionDataTask, 18
NSURLSessionTaskDelegate, 119
NSUserDefaults, 36, 37, 178
NSUUID, 234
NSXMLParser, 205–206

O

Objective-C, 13–25
  blocks
    declaring, 18
    exposing to JavaScript, 150–151
  categories, 22–23
  code structure, 15–17
  delightfulness of, 13
  garbage collection, 18
  implementation, declaring, 16–17
  ivars, 15–16
  message passing, 14–15
  private methods, lack thereof, 16
  property synthesis, 17
  reference counting, 18–19
odcctools, 78, 84
OpenSSH, 78
OpenSSL, 94, 228–229
openssl command, 138
openURL, 132–137
otool, 53, 78, 84–86,
  inspecting binaries, 90–92
output encoding, 200–202

P

p12 file, 113
parameterized SQL, 203–204
pasteboardDB file, 165–167

pasteboards, 164–169
  canPerformAction, 168
  disabling copy/paste, 168–169
  pasteboardDB file, 165–167
  pasteboardWithUniqueName, 165–167
  UISearchBar, 165
pasteboardWithUniqueName, 165–167
PhoneGap, 11, 150
physical attackers, 161
PIE (position-independent
      executable), 53–54
  removing, 87
plist files, 29–31
  converting, 30–31
  XML, 29–30
plutil command, 30–31
popen, 10
position-independent executable
      (PIE), 53–54
  removing, 87
predicate injection, 204–205
  LIKE operator, 205
  MATCHES operator, 205
  wildcards, 204–205
predicates, 205
predicateWithFormat, 204–205
Preferences directory, 36
printf command, 87, 190–192
print object command, 67–68
privacy concerns, 233–248
  advertisingTrackingEnabled, 235
  bluetooth low energy (BTLE), 244
  cookies, 237–238
  Do Not Track, 236–237
  geolocation, 238–240
    accuracy, 239
    CLLocationManager, 240
    locationManager, 244
    risks, 238–239
  GPS, 238
  HealthKit, 240–241
  iBeacons, 244–247
    CBPeripheralManager, 246
    CLBeaconRegion, 244–246
    CLLocationManager, 244
    startMonitoringForRegion, 244
  M7 processor, 242
  MAC address, 234
  microphone, 233

privacy policies, 247–248
proximity tracking, 244–247
requesting permission, 243
unique device identifier (UDID),
    233–235
  advertisingIdentifier, 235
  identifierForVendor, 234
  NSUUID, 234
  uniqueIdentifier, 234
private methods, 16
property synthesis, 17
protocols, 20–22
  declaring, 21–22
proximity tracking, 244–247
proxy setup, 43–50

Q

Quick Look, 35, 68
QuickType, 177

R

reference counting model, 18–19
  retain and release, 18–19
references, strong and weak, 19
release, 18–19
remote device wipe, 5, 6
removeAllCachedResponses, 75
respringing, 79, 101
restorationIdentifier, 184–185
retain, 18–19
return-to-libc attack, 8
RNCryptor, 186, 230
rootViewController, 183
rsync command, 78

S

safe string APIs, 195
Sandbox, 4–5
Saved Application State directory, 37
Seatbelt, 4–5
SecCertificateRef, 112–113
SecIdentityRef, 112–113
SecItemAdd, 186, 212, 215, 219
SecItemCopyMatching, 216
SecItemDelete, 216
SecItemUpdate, 215
SecRandomCopyBytes, 227
SecTrustRef, 112–113

Secure Boot, 4
SecureNSCoder, 186–187
securityd, 7
serialization, 21
setAllowsAnyHTTPSCertificate, 108
setJavaScriptCanOpenWindows-
    Automatically, 159
setJavaScriptEnabled, 159–160
setResourceValue, 188
setSecureTextEntry, 175–177
setShouldResolveExternal-
    Entities, 206
Shared directory, 37
sharedCredentialStorage, 120–122
sharedHTTPCookieStorage, 237
shared Keychains, 218–219
shouldAllowExtensionPoint-
    Identifier, 143
should messages, 20
shouldSaveApplicationState, 20
shouldStartLoadWithRequest, 148
sideloading, 77–80
signed integer, 196
signedness, 51, 196
Simulator, 43–46
    camera, 43
    case-sensitivity, 43
    installing certificates, 44
    Keychain, 43
    PBKDF2, 43
    proxying, 44–46
    trust store, 44
SpringBoard, 79
SQL injection, 201, 203–204
    parameterized SQL, 203–204
    SQLite, 203–204
SSH, 28, 82
SSL, *see* TLS (Transport Layer
    Security)
SSL Conservatory, 115–117
SSL Killswitch, 96–97
stack, 8, 53–54, 190, 193
startMonitoringForRegion, 244
state preservation, 184–187
    leaks, 184–185
    restorationIdentifier, 184–185
    secure, 185–187
static analysis, 54
std::string, 195

strcat, 195
strcpy, 194, 195
stringByEvaluatingJavaScriptFrom-
    String, 149–150
strlcat, 195–196
strlcpy, 195–196
strong references, 19
stunnel, 46
subclassing, 23
synthesize, 17
syslog, 162, 190

T

task_for_pid-allow, 82
tcpdump command, 78
tcpprox, 49–50
TCP proxying, 49–50
test devices, 42
text segment, 85–86
Theos, 97–98
thin binaries, 85
ThisDeviceOnly, 212
TLS (Transport Layer Security),
    108–119, 127–129
    BEAST attack, 118
    bypassing validation, 44–47, 119
    certificate pinning, 114–117,
        123–124
    CRIME attack, 118
    mutual authentication, 112–113
    setAllowsAnyHTTPSCertificate, 108
    validation, category bypasses, 22
tmp directory, 37, 80, 187
Today screen, 131
TOFU (trust on first use), 127
TouchID, 231–232
    LAContext, 231–232
Transport Layer Security, *see* TLS
    (Transport Layer Security)
Tribbles, 51
trust on first use (TOFU), 127
tweaks, Cydia Substrate, 97

U

UDID (unique device identifier),
    233–235
    advertisingIdentifier, 235
    identifierForVendor, 234

UDID (unique device identifier), *continued*
  NSUUID, 234
  uniqueIdentifier, 234
UIActivity, 139–140
UIAlertView, 183
UI Layers, 182–183
UIPasteBoard, 144, 164–169
UIPasteboardNameFind, 165
UIPasteboardNameGeneral, 165
UIRequiredDeviceCapabilities, 34
UIResponderStandardEditActions, 169
UISearchBar, 165, 175
UITextField, 175
UITextView, 175
UIView, 182–183
UIWebView, 200, 201
UIWindow, 182–183
unique device identifier (UDID),
    233–235
  advertisingIdentifier, 235
  identifierForVendor, 234
  NSUUID, 234
  uniqueIdentifier, 234
uniqueIdentifier, 234
universal links, 137–138
unsigned integer, 196
URL loading system, 107
  credential persistence types, 111
  HTTP basic authentication, 110–111
  HTTP redirects, 113–114
  NSURLConnection, 108
  NSURLConnectionDelegate, 109
  NSURLCredential, 120
  NSURLCredentialStorage, 110–111
  NSURLRequest, 108
  NSURLResponse, 108
  NSURLSession, 117–122
  NSURLSessionConfiguration, 117–119
  NSURLSessionTaskDelegate, 119
  sharedCredentialStorage, 120–122
  stored URL credentials, 121–122
URL schemes, 132–133
  CFBundleURLSchemes, 133
  defining, 132–133
  hijacking, 136–137
  validating URLs and senders, 134
USB, TCP proxying, 84
usbmuxd command, 84

user preferences, 178
UUID, 27
uuidgen, 244

V

Valgrind, 55
vbindiff command, 78, 88
vfork, 10
.vimrc file, 30
vmaddr, 88

W

wardriving, 238
warning policies, 51
watchdog, 58–59
watchmedo command, 58–59
weak_classdump, 93
weak references, 19
web apps, 147–160
WebViews, 9, 147–160
  Cordova, 154–157
    risks, 156
    XmlHttpRequest, 155
  JavaScript, 149
    executing in Cordova, 154–157
    executing in UIWebView, 149–150
    stringByEvaluatingJavaScriptFrom-
      String, 149–150
  JavaScript–Cocoa bridging, 150–157
  JavaScriptCore, 149–154
    blocks, 150–151
    JSContext, 152–154
    JSExport, 151–152
  just-in-time (JIT) compiler, 149
  Nitro, 148, 149
  NSURLRequest, 148–149
  UIWebView, 147–150
  WebKit, 11, 147–148
  WKWebView, *see* WKWebView
whitelisting, 149, 152, 200–201
will messages, 20
willSendRequestForAuthentication-
    Challenge, 112
Wireshark, 46
WKPreferences, 160
WKWebView, 148 158–160
  addUserScript, 159
  benefits of, 159–160

didFinishNavigation, 159–160
hasOnlySecureContent, 159–160
setJavaScriptCanOpenWindows-
    Automatically, 159
setJavaScriptEnabled, 159–160
WKPreferences, 160
WKUserScript, 159
WKWebViewConfiguration, 160

X

xcodebuild, 190
Xcode setup, 50–53, 55
   warnings, 51–53
Xcon, 10
XML injection, 207
   NSXMLParser, 205–206
   XML external entities, 205–206
   XPath, 207
XN (eXecute Never), 8–9
XPath, 207
XSS (cross-site scripting), 199–200
   input sanitization, 200–201
   output encoding, 201–202
xxd command, 88

The fonts used in *iOS Application Security* are New Baskerville, Futura, The Sans Mono Condensed and Dogma. The book was typeset with $\LaTeX\,2_\varepsilon$ package nostarch by Boris Veytsman *(2008/06/06 v1.3 Typesetting books for No Starch Press)*.

The book was printed and bound by Sheridan Books, Inc. in Chelsea, Michigan. The paper is 60# Finch Offset, which is certified by the Forestry Stewardship Council (FSC). The book uses a layflat binding, in which the pages are bound together with a cold-set, flexible glue, and the first and last pages of the resulting book block are attached to the cover. The cover is not actually glued to the book's spine, and when open, the book lies flat and the spine doesn't crack.

The Electronic Frontier Foundation (EFF) is the leading organization defending civil liberties in the digital world. We defend free speech on the Internet, fight illegal surveillance, promote the rights of innovators to develop new digital technologies, and work to ensure that the rights and freedoms we enjoy are enhanced — rather than eroded — as our use of technology grows.

EFF.ORG
ELECTRONIC FRONTIER FOUNDATION
Protecting Rights and Promoting Freedom on the Electronic Frontier

UPDATES

Visit *https://www.nostarch.com/iossecurity* for updates, errata, and other information.

More no-nonsense books from **NO STARCH PRESS**

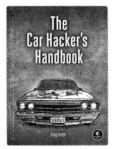

THE CAR HACKER'S HANDBOOK
by CRAIG SMITH
SPRING 2016, 352 PP., $49.95
ISBN 978-1-59327-703-1

BLACK HAT PYTHON
Python Programming for Hackers and Pentesters
by JUSTIN SEITZ
DECEMBER 2014, 192 PP., $34.95
ISBN 978-1-59327-590-7

GAME HACKING
Developing Autonomous Bots for Online Games
by NICK CANO
SPRING 2016, 384 PP., $44.95
ISBN 978-1-59327-669-0

ROOTKITS AND BOOTKITS
Reversing Modern Malware and Next Generation Threats
by ALEX MATROSOV, EUGENE RODIONOV, *and* SERGEY BRATUS
SPRING 2016, 304 PP., $49.95
ISBN 978-1-59327-716-1

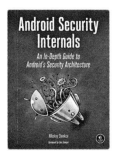

ANDROID SECURITY INTERNALS
An In-Depth Guide to Android's Security Architecture
by NIKOLAY ELENKOV
OCTOBER 2014, 432 PP., $49.95
ISBN 978-1-59327-581-5

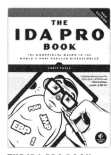

THE IDA PRO BOOK, 2ND EDITION
The Unofficial Guide to the World's Most Popular Disassembler
by CHRIS EAGLE
JULY 2011, 672 PP., $69.95
ISBN 978-1-59327-289-0

PHONE:
800.420.7240 OR
415.863.9900

EMAIL:
SALES@NOSTARCH.COM
WEB:
WWW.NOSTARCH.COM